Distributed Kill Chains

Drawing Insights for Mosaic Warfare from the Immune System and from the Navy

NICHOLAS A. O'DONOUGHUE, SAMANTHA MCBIRNEY,
BRIAN PERSONS

Prepared for the Defense Advanced Research Projects Agency
Approved for public release; distribution unlimited

RAND NATIONAL DEFENSE RESEARCH INSTITUTE

For more information on this publication, visit www.rand.org/t/RRA573-1

Library of Congress Cataloging-in-Publication Data is available for this publication.
ISBN: 978-1-9774-0613-2

Published by the RAND Corporation, Santa Monica, Calif.

Cover: F-35: Michael Fitzsimmons/Getty Images;
T-cells: Spectral-Design/Getty Images/iStockphoto

Cover design: Rick Penn-Kraus

Preface

Mosaic warfare is a concept that the Strategic Technology Office of the Defense Advanced Research Projects Agency (DARPA) is developing to acquire, field, and employ forces in a manner that is radically different from what is currently done in the U.S. Department of Defense. Like the ceramic tiles in mosaics, individual warfighting platforms are put together to make a larger picture, known as a *force package*.

In 2019, DARPA asked the RAND Corporation's National Security Research Division (NSRD) to explore and validate the fundamental value propositions of the three key Mosaic warfare architectural attributes by means of modeling and simulation. The study also aimed to identify near-term mosaic experimentation opportunities and hypotheses through functional decomposition and recomposition of existing systems and architectures. The results of this research project, *Mosaic Warfare Experimentation Architecture Study*, are documented in this report and two others:

- Justin Grana, Jonathan Lamb, and Nicholas A. O'Donoughue, *Findings on Mosaic Warfare from a Colonel Blotto Game*, Santa Monica, Calif.: RAND Corporation, RR-4397-OSD, 2021.
- Timothy R. Gulden, Jonathan Lamb, Jeff Hagen, and Nicholas A. O'Donoughue, *Modeling Rapidly Composable, Heterogeneous, and Fractionated Forces: Findings on Mosaic Warfare from an Agent-Based Model*, Santa Monica, Calif.: RAND Corporation, RR-4396-OSD, 2021.

These reports will be of interest to military acquisition specialists and strategists. This report in particular is focused on identifying potential challenges in the transition to Mosaic warfare and changes that might be required to cope with its expected traits. To better understand these challenges and the best path forward, the authors have conducted a pair of case studies: (1) an analysis of the human immune system's response to pathogens and (2) an analysis of the Naval Integrated Fire Control—Counter Air (NIFC-CA) project. The report includes rudimentary biology discussions to support the use of the human immune system as an analogy to Mosaic warfare.

This research was sponsored by DARPA and conducted within the Acquisition and Technology Policy Center of the RAND National Security Research Division (NSRD), which operates the National Defense Research Institute (NDRI), a federally funded research and development center sponsored by the Office of the Secretary of Defense, the Joint Staff, the Unified Combatant Commands, the Navy, the Marine Corps, the defense agencies, and the defense intelligence enterprise.

For more information on the RAND Acquisition and Technology Policy Center, see www.rand.org/nsrd/atp or contact the director (contact information is provided on the webpage).

Contents

Figures

Tables

Summary

As the U.S. Department of Defense (DoD) continues to innovate, several new approaches to fighting conflicts are under consideration to reduce cost and increase effectiveness and resiliency in a variety of scenarios. These approaches have emerged as complements, or even alternatives, to a more traditional focus on high-capability, high-cost platforms, such as the F-35 fighter, B-21 bomber, or *Ford*-class aircraft carrier. One new approach, supported by the U.S. Army and the U.S. Air Force, is known as *multidomain operations* and looks to build non-linear *kill webs* by rapidly bringing together sensors and shooters across service, domain, and functional stovepipes.[1]

Mosaic warfare is a concept that the Strategic Technology Office (STO) of the Defense Advanced Research Projects Agency (DARPA) is developing to acquire, field, and employ forces in a radically different manner from what is currently done in DoD. Like the ceramic tiles in mosaics, individual warfighting platforms are assembled to make a larger picture or, in this case, a force package.

Achieving a Mosaic warfare design will likely require an evolutionary journey rather than a radical revolutionary approach that happens overnight. Figure S.1 lays out four developmental stages that DARPA envisions on the road to Mosaic warfare, along with their challenges.

[1] For example, see remarks by ADM Philip Davidson, commander of U.S. Indo-Pacific Command, to the Senate Armed Services Committee. Davidson said that "the U.S. government must continue to pursue multi-domain capabilities to counter anti-air capabilities . . . ." (Philip S. Davidson, *Statement of Admiral Philip S. Davidson, U.S. Navy Commander, U.S. Indo-Pacific Command Before the Senate Armed Services Committee on U.S. Indo-Pacific Command Posture*, testimony presented before the U.S. Senate Committee on Armed Services, Washington, D.C., February 12, 2019.)

Figure S.1
The Pathway to Mosaic Warfare

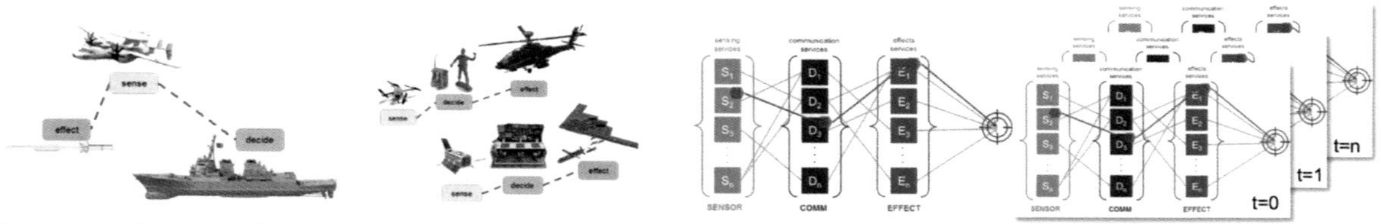

	Distributed kill chain	System of systems	Adaptive kill web	Mosaic warfare
Example	NIFC-CA	SoSITE	TBD	TBD
Description	Manual integration of existing systems	Systems prepped for multiple battle configurations	Semiautomated ability to select a predefined effects web prior to mission	Ability to compose new effects webs at campaign time
Benefits	• Extends effective range • Increases engagement opportunity	• Enables faster integration and more-diverse kill chains	• Allows pre-mission adaptation • More lethal, imposes complexity on adversary	• Adaptable to dynamic threat and environment • Scalable to many simultaneous engagements
Challenges	• Static • Long to build • Difficult to operate and scale	• Each architecture static • Limited ability to adapt • Cannot add new capabilities on the fly • Difficult to operate and scale	• Static "playbook" • Limited number of kill chains • Might not scale well	• Scale limited by human decisionmakers

SOURCE: Adapted from Timothy Grayson, "Mosaic Warfare," briefing slides, Washington, D.C.: Defense Advanced Research Projects Agency, Strategic Technology Office, July 27, 2018.
NOTE: NIFC-CA = Naval Integrated Fire Control—Counter Air; SoSITE = System of Systems Integration Technology and Experimentation.

As these challenges are overcome, the benefits identified at each stage will be realized. Although the technological stages should follow an evolutionary approach, the way that systems are procured, evaluated, and fielded might require radical changes to cope with the expected traits of Mosaic warfare, such as the dynamic composition of capabilities in response to ever-changing and unexpected environments.

To better understand these challenges and the best path forward, we have conducted a pair of case studies: (1) an analysis of the human immune system's response to pathogens to draw insights about the requirements, challenges, and opportunities of Mosaic warfare and (2) an analysis of the Naval Integrated Fire Control—Counter Air (NIFC-CA) project to understand the decisions made and lessons learned in its development and fielding.[2]

Case Study Selection

DARPA asked the RAND Corporation's National Security Research Division to conduct a case study to understand the potential challenges facing the transition to Mosaic warfare. The first step was to highlight potential programs to study and compare them, specifically looking at six principal features of a Mosaic warfare system: *fractionation* (segregation of warfighting functions onto large numbers of smaller platforms, as opposed to small numbers of integrated platforms); *heterogeneity* (the system comprises different parts with unique capabilities); *rapid composability* (response actions are tailored and made up of differing parts, depending on the threat); system architecture; scalability and multiagent collaboration; and artificial intelligence/machine learning (AI/ML) and autonomy.

From a list of seven potential case studies, we ultimately settled on NIFC-CA as the system of choice given its similarity to Mosaic warfare, the availability of data on its development and acquisition cycle,

[2] *NIFC-CA* is typically pronounced *niff-cah*.

and the fact that the system has achieved initial operational capability (IOC) for some components.[3]

During the process of discussing possible case study programs, we noted that the natural world has many systems that share attributes with the Mosaic warfare vision, including disaggregation of capabilities across several independently operating platforms. We conducted a similar survey of potential biological analogues and settled on the immune system and its response to pathogens as a second case study, largely because of the shared traits of heterogeneity and composability.

Lessons from the Immune Response

Figure S.2 shows a simplified illustration of the immune system's response to pathogens, broken into two stages called *innate immunity* and *adaptive immunity*. We have chosen to use simplistic descriptions of the immune response for the sake of brevity. Although the body of our report contains additional details, a full treatment of the immune system's many components is beyond the scope of this work.

The innate immune response provides the first line of defense against invading pathogens, behaving similarly regardless of the specific pathogen encountered. Several components constitute the innate immune system, the most significant being physical barriers.[4] Upon pathogen invasion, myeloid cells are recruited to the site of infection. Several subtypes of *myeloid cells* engulf pathogenic microbes and destroy them, while other subtypes participate in acute inflammatory and allergic responses.[5] Constantly renewed to reinforce strength, the innate immune response successfully wards off most incursions. However, the innate system cannot always recognize or eliminate infec-

---

[3] For example, the Navy declared NIFC-CA IOC for the E-2D Advanced Hawkeye in 2014. See Dave Majumdar, "Navy Declares IOC for E-2D Advanced Hawkeye," USNI News, October 16, 2014.

[4] David D. Chaplin, "Overview of the Immune Response," *Journal of Allergy and Clinical Immunology*, Vol. 125, No. 2, Supp. 2, February 2010.

[5] Hiroshi Kawamoto, and Nagahiro Minato, "Myeloid Cells," *International Journal of Biochemistry & Cell Biology*, Vol. 36, No. 8, August 2004.

Figure S.2
Simplified Model of Immune Response

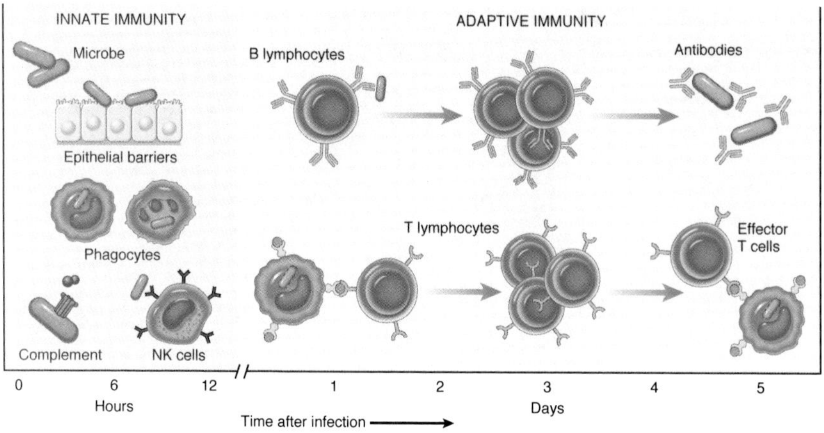

SOURCE: Vinay Kumar, Abul K. Abbas, and Nelson Fausto, eds., *Robbins and Cotran Pathologic Basis of Disease*, 7th ed., Philadelphia, Pa.: Saunders, 2004. Used with permission.

tious organisms, at which point the second line of defense, the adaptive immune system, enters the battlefield.[6]

An intact adaptive immune response includes contributions from large subpopulations of *lymphocytes*, each population having a unique morphology and playing a distinct role. Mature lymphocytes have three main lineages—B cells, T cells, and natural killer (NK) cells—with B cells and T cells being the primary components of the adaptive immune response.[7] B cells are primarily responsible for *antibody-driven* immunity, producing antibodies that travel throughout the body in the bloodstream, bind to antigens, and subsequently neutralize them.[8] In contrast, T cells destroy cells that have already been infected, prevent-

[6] Charles A. Janeway, Jr., Paul Travers, Mark Walport, and Mark J. Shlomchik, *Immunobiology*, 5th ed., New York: Garland Science, 2001.

[7] Kawamoto and Minato, 2004.

[8] William Hoffman, Fadi G. Lakkis, and Geetha Chalasani, "B Cells, Antibodies, and More," *Clinical Journal of the American Society of Nephrology*, Vol. 11, No. 1, January 2016.

ing further spread of the disease.[9] Each T cell bears receptors to a single specific antigen, so a repertoire of T cells that can protect against a multitude of pathogens must include a very large number of cells. NK cells serve to contain viral infection while the adaptive immune response is generating antigen-specific T cells that can clear the infection.[10]

As part of our analysis, we draw a series of analogies between the immune system and the Mosaic warfare vision:

- We cast the immune response to pathogens in the context of
  - a U.S. Air Force *kill chain* used to detect, maintain custody of, and engage targets in a dynamic scenario.
  - joint U.S. doctrine on *base defense*, wherein various resources and methods are used to effectively repel a variety of attacks on an installation.
  - rapid mobilization of forces in preparation for hostilities.
- We cast vaccines in the context of strategic preparations (intelligence gathering) for conflict.
- We cast the failures of the immune system, such as allergic reactions and overactivity, autoimmune disorders, or deficiencies, in the context of military failures, such as fratricide or cyberspace operations, that interfere with coordination of the mosaic architecture.

From these analogies, we observe that a Mosaic warfare system, insofar as it obeys our analogy to the immune system, could be adaptable and resilient to failures of individual platforms. Mosaic warfare might also be inefficient, relying on an abundance of low-cost systems that could present a significant challenge for centralized command-and-control (C2) schemes.

We note from this analogy that a successful Mosaic warfare system will require robust communication to enable distributed col-

---

9 Mads Hald Andersen, David Schrama, Per thor Straten, and Jürgen C. Becker, "Cytotoxic T Cells," *Journal of Investigative Dermatology*, Vol. 126, No. 1, January 2006.

10 Eric Vivier, David H. Raulet, Alessandro Moretta, Michael A. Caligiuri, Laurence Zitvogel, Lewis L. Lanier, Wayne M. Yokoyama, and Sophie Ugolini, "Innate or Adaptive Immunity? The Example of Natural Killer Cells," *Science*, Vol. 331, No. 6013, January 2011.

lection, processing, and dissemination of information, as well as distributed C2 schemes. Rapid, tailored responses to a wide array of possible threats require the forward staging of a large force, improvements in rapid mobilization and deployment of reserve forces, or distributed rapid-production facilities.

Lessons from the Navy's Naval Integrated Fire Control—Counter Air Project

We present a case study of the U.S. Navy's NIFC-CA project, draw lessons learned from the development of NIFC-CA, and apply these lessons, where possible, to the Mosaic warfare concept.[11] The focus of the NIFC-CA case study is to analyze the organizational construct, the experimentation process, and the documentation used to support NIFC-CA decisionmaking and budgets.

Since 1996, the NIFC-CA project has been developing a family-of-systems (FoS) capability to defeat overland cruise missiles and other over-the-horizon air warfare threats. An illustration of a recent test integrating the F-35 into the NIFC-CA architectures is shown in Figure S.3.

The original NIFC-CA concept consisted of five pillar programs for its *From-the-Sea* kill chain: the E-2D Advanced Hawkeye and the Joint Land Attack Cruise Missile Defense Elevated Netted Sensor System (JLENS) to provide long-range detection of incoming cruise missiles, the Cooperative Engagement Capability network to provide dissemination of detections and tracking data, and the AEGIS fire control system and Standard Missile (SM) 6 interceptor to provide engage-

---

[11] We use NIFC-CA as a relevant case study because of its similarity to the Mosaic warfare vision. Indeed, this analogy also highlights that many of the Mosaic warfare concepts are not new. In fact, the original ideals of net-centric warfare align closely with Mosaic warfare's goals regarding ubiquitous communications, coordination, and adaptive kill chains. See Clay Wilson, *Network Centric Operations: Background and Oversight Issues for Congress*, Washington, D.C.: Congressional Research Service, March 15, 2007.

Figure S.3
Graphical Illustration of a Naval Integrated Fire Control—Counter Air Engagement

SOURCE: Commander, Naval Surface Force, U.S. Pacific Fleet, "Navy and Marines Demonstrate Integrated Fire Control," webpage, undated.
NOTE: In this engagement, the F-35 provides detection of an incoming threat and transmits targeting information to a ground station (AEGIS Ashore) that cues the launch of an interceptor missile and guides it to the threat.

ment of threats.[12] NIFC-CA was structured as an integration project rather than as a program of record and relied heavily on its pillar programs to provide test and integration events.

---

12 Similar kill chains were devised for interceptor launch *From-the-Air* and *From-the-Land*, but *From-the-Sea* was developed first.

From our analysis of the organization, acquisition approach, and experimentation of NIFC-CA, we draw several lessons for Mosaic warfare:

- The U.S. Navy's decision to structure NIFC-CA as a project with minimal (approximately $35 million per year) research and development (R&D) and test and evaluation (T&E) funding for integration and interoperability of the pillar programs rather than to form a separate acquisition program with significant procurement funding resulted in a small increase in funding requirements on the underlying pillar programs of record. However, this decision also allowed NIFC-CA to survive the turbulence of budget cycles and reviews, including the cancellation of one of its core pillars (JLENS).
- NIFC-CA began with three exemplar kill chains and identified a standard for data structures early. This agreed-upon format and early concept definition allowed the early use of experiments and ensured that component programs were all capable of understanding a common data format.
- NIFC-CA leveraged underlying exercises for the component programs, adding a small number of well-defined internally funded test events. Thus, NIFC-CA used the testing infrastructure of the pillar programs rather than having to depend on the project office to create its own.
- Instead of a rigorous T&E master plan, which is typically required for major acquisition programs to test against predefined requirements, NIFC-CA's test events were approached in a more scientific manner designed to test hypotheses. Test failure was seen as a learning opportunity rather than a project setback.
- NIFC-CA's lengthy development cycle was influenced by the acquisition timeline of its pillar programs, specifically the fact that many of the required capabilities were not fielded at the inception of NIFC-CA and were developed in parallel with the NIFC-CA architecture. The length of the development cycle was driven primarily by the need to integrate the legacy pillar programs into an integrated FoS, which they were not originally envisioned to sup-

port. Future increments of NIFC-CA will not require such long development cycles and are planned to be accomplished within much shorter timelines.

Recommendations

We analyzed these findings in the context of Mosaic warfare and arrived at a series of recommendations that span three classes: programmatic, R&D, and T&E. Further justification is provided in the main body of our report, but we summarize the recommendations in this section.

Programmatic

- Identify pathfinder programs and exemplar vignettes against which to develop the Mosaic warfare architecture.
- Structure acquisition of the Mosaic warfare architecture as a project with "pillar" programs rather than as a program of record.

Research and Development

- Develop algorithms for robust data sharing and distributed processing across large networks of mobile systems.
- Develop low-cost rapid manufacturing of platforms and payloads that will make up the Mosaic warfare system.
- Develop distributed manufacturing capabilities to construct the platforms and payloads that will make up the Mosaic warfare system close to where they are needed.
- Develop distributed and automated C2 algorithms and processes that can translate high-level commands to individual platforms and adapt behavior to changing environments and threat behavior.

Test and Evaluation

- Rely on pillar programs to provide early test opportunities, and approach each test event with a willingness to fail and a learning objective.
- Develop an approach to test adaptable systems (including AI/ML algorithms).
- Test the resilience of the Mosaic warfare architecture to failure or compromise of component systems.

Limitations in Our Analysis

The primary limitation in the immune response analogy is that the immune response is inherently a defensive scenario in which numerical superiority and proximity of forces are advantages. This model applies directly to base-defense scenarios but somewhat stresses the application to force projection. Our comparisons, as drawn, are agnostic to these facts, but the underlying success of the immune system depends on the ability to muster large numbers of both general and specialized response forces and the ability to communicate information to reserve forces regarding the location and composition of adversary forces.

The primary limitation in our use of NIFC-CA as a case study is the fact that NIFC-CA, while distributed, lacks the dynamic kill chain aspects that are central to Mosaic warfare. In the development and testing of NIFC-CA, tremendous care was taken to define and test the kill chains (from the E-2D Hawkeye or the AEGIS sensor to the F/A-18 strike aircraft or the SM-6). Mosaic warfare, by contrast, is defined by the presence of multiple options for each element of the kill chain and the promise that some elements can be defined later and integrated into the architecture. This characteristic presents significant challenges for acquisition of the new systems and integration into the architecture.

Acknowledgments

We would like to express our sincere thanks to Timothy Grayson, director of DARPA's STO, and Lt Col Daniel Javorsek, U.S. Air Force, a program manager in STO and our action officer, for their insight, direction, and support. We would also like to extend our appreciation to Ronald Hill and David Ott, support contractors for Grayson and Javorsek, respectively, for their endless assistance in coordinating the details of this project. We wish to thank Samuel Earp and John Kamp, who provide subject-matter expertise to STO and were instrumental in strengthening our analysis.

Our report evolved greatly during the review and quality assurance process. We wish to thank Chris Carson, John Yurchak, Ritika Chaturvedi, David Relman, Jim Powers, and Joel Predd for their assistance with reviewing the documents and Sydne Newberry for editing the document and improving our ability to coherently arrange our findings.

Abbreviations

ACB	Advanced Capability Build
AI	artificial intelligence
AIDS	acquired immunodeficiency syndrome
AI/ML	artificial intelligence/machine learning
AMOD	AEGIS Modernization
APC	antigen-presenting cell
C2	command and control
CEC	Cooperative Engagement Capability
CEM	cooperative engagement mode
CONOP	concept of operations
CSG	carrier strike group
DARPA	Defense Advanced Research Projects Agency
DoD	U.S. Department of Defense
DODAF	DoD Architecture Framework
FoS	family of systems

FTS	From-the-Sea
FY	fiscal year
HIV	human immunodeficiency virus
IFC	Integrated Fire Control
IOC	initial operational capability
IPT	Integrated Product Team
JLENS	Joint Land Attack Cruise Missile Defense Elevated Netted Sensor System
JP	Joint Publication
JSF	Joint Strike Fighter
LFT	Live-Fire Test
NIFC-CA	Naval Integrated Fire Control—Counter Air
NK	natural killer
OCMD	overland cruise missile defense
OTH	over-the-horizon
PEO IWS	Program Executive Officer for Integrated Warfare Systems
R&D	research and development
RDT&E	research, development, test, and evaluation
SEI&T	System Engineering, Integration and Test
SM	Standard Missile
SoS	system of systems

SoSITE	System of Systems Integration Technology and Experimentation
STO	Strategy Technology Office
T&E	test and evaluation
TEMP	test and evaluation master plan
UARC	university-affiliated research center
WWII	World War II

CHAPTER ONE

Introduction

As the U.S. Department of Defense (DoD) continues to innovate, several new approaches to fighting conflicts are under consideration to reduce cost and increase effectiveness and resiliency in a variety of scenarios. These approaches have emerged as complements, or even alternatives, to a more traditional focus on high-capability, high-cost platforms, such as the F-35 fighter, B-21 bomber, or *Ford*-class aircraft carrier. One new approach, supported by the U.S. Army and the U.S. Air Force, is known as *multidomain operations* and looks to build non-linear *kill webs* by rapidly bringing together sensors and shooters across service, domain, and functional stovepipes.[1]

Another novel concept under development by the Strategic Technology Office of the Defense Advanced Research Projects Agency (DARPA) is *Mosaic warfare*. Like the ceramic tiles that compose mosaics, DARPA's Mosaic warfare concept assembles individual warfighting platforms to make a larger picture or, in this case, a force package.[2] As stated in a DARPA-commissioned report, *Restoring America's*

---

[1] For example, see remarks by ADM Philip Davidson, commander of U.S. Indo-Pacific Command, to the Senate Armed Services Committee. Davidson said that "the U.S. government must continue to pursue multi-domain capabilities to counter anti-air capabilities . . . ." (Philip S. Davidson, *Statement of Admiral Philip S. Davidson, U.S. Navy Commander, U.S. Indo-Pacific Command Before the Senate Armed Services Committee on U.S. Indo-Pacific Command Posture*, testimony presented before the U.S. Senate Committee on Armed Services, Washington, D.C., February 12, 2019.)

[2] For more detail, see Defense Advanced Research Projects Agency, "DARPA Tiles Together a Vision of Mosaic Warfare," webpage, undated.

Military Competitiveness: Mosaic Warfare, by the Mitchell Institute for Aerospace Studies,

> "Mosaic" is a force design concept for a systems warfare strategy. The concept is designed to address the demands of the future strategic environment and the shortcomings of the current force. Mosaic warfare exploits both the ability of advanced networks to seamlessly share information across an area of operations and recent developments in processing, computing, and networking. Functional capabilities, such as radar, fire control, and missiles, that once had to be hosted on a common platform, like a sophisticated combat aircraft, can now be disaggregated into their smallest practical elements. In the mosaic concept, platforms are "decomposed" into their smallest practical functions, creating collaborative "nodes" in a networked kill web that is highly resilient and can remain operationally effective, even as an adversary attrits some of the web's elements.[3]

The Mitchell Report also postulates that migrating to a mosaic force design will require investment and oversight in the following areas:

- automated technology that can share information across different security levels . . .
- appropriate policy for test, validation, and verification of artificial intelligence . . .
- multiple and complimentary approaches to spoof-proofing artificial intelligence . . .
- [maintenance of] current force structure and programs of record . . .

---

[3] David A. Deptula and Heather R. Penney, with Lawrence A. Stutzriem and Mark A. Gunzinger, *Restoring America's Military Competitiveness: Mosaic Warfare*, Arlington, Va.: Mitchell Institute for Aerospace Studies, September 2019, pp. 3–4. According to Deptula et al., 2019, p. 7, *systems warfare* is a "theory of warfare that does not rely on attrition or maneuver to achieve advantage and victory over the adversary. Instead, systems warfare targets critical points in an adversary's system to collapse its functionality and render it unable to prosecute attack or defend itself. A major objective of this approach is to maximize desired strategic returns per application of force (achieve best value)."

- operationally focused cost assessment of force design alternatives . . .
- [aggressive investment] in developing and fielding mosaic enablers . . .
- [experimentation] with mosaic operational concepts, architectures, and empowered command and control [C2] at the edge.[4]

DARPA's Path Forward

Achieving a Mosaic warfare design will likely require an evolutionary journey rather than a radical revolutionary approach that happens overnight. In Figure 1.1, DARPA's vision of a pathway to a Mosaic warfare design includes four stages:

- a distributed kill chain used by the Naval Integrated Fire Control—Counter Air (NIFC-CA)[5] project
- system-of-systems (SoS) architecture employed by the System of Systems Integration Technology and Experimentation (SoSITE) program[6]
- a futuristic adaptive kill web
- Mosaic warfare.

Achieving a Mosaic warfare design will require addressing the challenges outlined in Figure 1.1. As these challenges are overcome, the benefits identified at each stage will be realized. Although the technological stages should follow an evolutionary approach, the way that systems are procured, evaluated, and fielded might require radical

---

4 Deptula et al., 2019, pp. 4–5.

5 *NIFC-CA* is typically pronounced *niff-cah.*

6 The SoSITE program, managed by DARPA, "seeks to develop and deliver systems architecture concepts for rapid integration of new U.S. technologies as they are developed, without requiring significant re-engineering of existing capabilities, systems, or systems of systems." (Jimmy Jones, "System of Systems Integration Technology and Experimentation (SoSITE)," webpage, Defense Advanced Research Projects Agency, undated.)

Figure 1.1
The Pathway to Mosaic Warfare

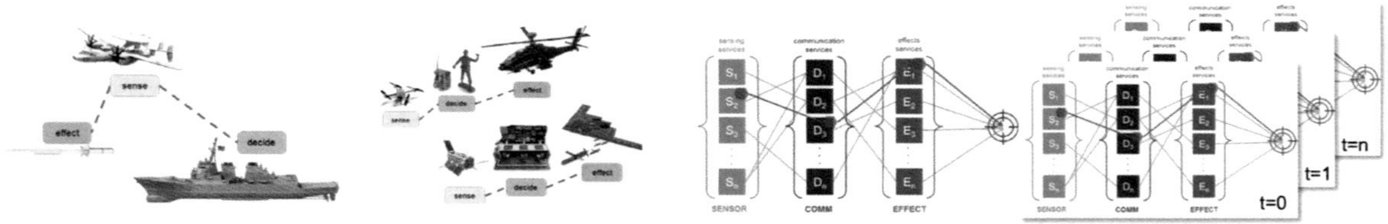

	Distributed kill chain	System of systems	Adaptive kill web	Mosaic warfare
Example	NIFC-CA	SoSITE	TBD	TBD
Description	Manual integration of existing systems	Systems prepped for multiple battle configurations	Semiautomated ability to select a predefined effects web prior to mission	Ability to compose new effects webs at campaign time
Benefits	• Extends effective range • Increases engagement opportunity	• Enables faster integration and more-diverse kill chains	• Allows pre-mission adaptation • More lethal, imposes complexity on adversary	• Adaptable to dynamic threat and environment • Scalable to many simultaneous engagements
Challenges	• Static • Long to build • Difficult to operate and scale	• Each architecture static • Limited ability to adapt • Cannot add new capabilities on the fly • Difficult to operate and scale	• Static "playbook" • Limited number of kill chains • Might not scale well	• Scale limited by human decisionmakers

SOURCE: Adapted from Timothy Grayson, "Mosaic Warfare," briefing slides, Washington, D.C.: Defense Advanced Research Projects Agency, Strategic Technology Office, July 27, 2018.
NOTE: To better understand these challenges and the best path forward, we have conducted a pair of case studies: (1) an analysis of the human immune system's response to pathogens to draw insights about the requirements, challenges, and opportunities of a mosaic system and (2) an analysis of the NIFC-CA project to understand the decisions made and lessons learned in its development and fielding.

changes to cope with the expected traits of Mosaic warfare, such as the dynamic composition of capabilities in response to constantly changing and unexpected environments.

We chose to study the immune system because it demonstrates heterogeneity (the immune system is made up of different parts with unique capabilities) and composability (response actions are tailored and made up of differing parts, depending on the threat). We chose to study NIFC-CA because of its similarity to Mosaic warfare and because the system has achieved initial operational capability (IOC) for some components.[7] We discuss these decisions in greater detail in Chapter Two.

Organization of This Report

The remainder of our report is organized into four chapters. In Chapter Two, we discuss the selection process for the two case studies presented and the analysis approach for each. Chapter Three details our immune response case study, arriving at a series of observations regarding desired and expected traits for a mosaic system and the potential benefits and pitfalls inherent in such an approach. Chapter Four details our NIFC-CA case study, arriving at several observations regarding a possible approach for acquiring Mosaic warfare systems from a programmatic perspective. Finally, Chapter Five organizes the observations of the two case studies into a series of recommendations grouped into three categories: (1) programmatic recommendations, (2) research and development (R&D) recommendations, and (3) test and evaluation (T&E) recommendations.

---

[7] For example, the Navy declared NIFC-CA IOC for the E-2D Advanced Hawkeye in 2014. See Dave Majumdar, "Navy Declares IOC for E-2D Advanced Hawkeye," USNI News, October 16, 2014.

CHAPTER TWO

Research Approach

This chapter details how we selected the two case studies for this project and describes how each study was conducted.

Approach and Objectives

In our report, we outline two related case studies. The first is an analogical comparison to the human immune response to pathogens. We draw analogues between various warfighting scenarios and behaviors and those of the immune system as it fights off pathogens, with the intention of highlighting characteristics and capabilities of the immune response that align with the envisioned attributes of Mosaic warfare. Through this case study, we identify traits of a Mosaic warfare system that will be beneficial, as well as those that present challenges that must be overcome.

The second case study is a comparison of Mosaic warfare to the NIFC-CA project. We analyze the programmatic decisions and governance of the NIFC-CA project and attempt to draw insights into how a Mosaic warfare acquisition should be structured and what challenges it is likely to face.

The rest of this chapter details how we arrived at the two case studies presented and our research methodology for each.

Case Study Selection

DARPA asked the RAND Corporation's National Security Research Division to conduct a case study to understand the potential challenges facing the transition to Mosaic warfare. The first step was to highlight several potential programs to study and compare, specifically looking at the six principal features of a Mosaic warfare system outlined in Table 2.1: fractionation, heterogeneity, rapid composability, system architecture, scalability and multiagent collaboration, and artificial intelligence/machine learning (AI/ML) and autonomy.

The list of systems we considered included military acquisition programs, such as the U.S. Navy's Littoral Combat Ship Mis-

Table 2.1
Mosaic Warfare Envisioned Attributes

Dimension	Mosaic Warfare Goals
Fractionation	Many platforms and platform types linked together in common data networks
Heterogeneity	Diverse number of multiservice/multiagency platforms included as needed
Rapid composability	Rapid process for introducing new effects or capabilities, certification unclear [a]
System architecture	Complex and rapidly evolving multidimensional relationships
Scalability and multiagent collaboration	Joint multidomain assets that interoperate and scale easily up and down as necessary
AI/ML and autonomy	Significant reliance on AI/ML and automation likely required [b]

NOTES:

[a] The interaction of new capabilities or effects with existing subsystems must be tested and certified to behave as expected or at least not violate legal and safety requirements. This is particularly challenging in Mosaic warfare because the individual subsystems might be applied to larger tasks or missions for which they were not originally intended.

[b] The apparent need for advanced AI/ML has significant policy implications, particularly if a Mosaic warfare construct is to be given the ability to employ lethal weapons. It might be necessary, particularly for early iterations, to develop a hybrid solution that integrates manned platforms such that the ability to employ weapons is not given to an AI/ML algorithm. This discussion is outside the scope of our report but is a critical point to understand as a strategy is developed to acquire Mosaic warfare systems.

sion Module architecture,[1] the U.S. Army's Future Combat Systems program,[2] and the U.S. Air Force's Advanced Battle Management System,[3] and commercial systems, such as microservices architectures and heterogeneous computing system-on-a-chip architectures.[4] We presented a list of these options, along with brief analysis of their suitability to the Mosaic warfare vision, to the sponsor and settled on the Navy's NIFC-CA project as the ideal candidate.

During the process of discussing possible case study programs, we noted that the natural world has many systems that share attributes with the Mosaic warfare vision, including disaggregation of capabilities across several independently operating platforms. This sharing of technological attributes is not, in itself, a revelation—biomimicry is a branch of technology development that recognizes the elegance and efficiency of natural processes and uses them as inspiration to solve complex human problems.[5] We conducted a brief survey of potential natural world analogues, which included swarming behavior of insects and birds, as well as the immune system's response to pathogens. We analyzed each case to determine whether it exhibited heterogeneity, loose coupling of control among elements, dynamic behavior, the ability to repurpose agents to a new task, localized decisionmaking, memory and learning, resistance or protection against threats, and

---

1 Program Executive Office Unmanned and Small Combatants, "LCS Mission Modules Program: Transition from Development to Production," briefing slides, Washington, D.C.: Naval Sea Systems Command, undated.

2 Christopher G. Pernin, Elliot Axelband, Jeffrey A. Drezner, Brian B. Dille, John Gordon IV, Bruce J. Held, K. Scott McMahon, Walter L. Perry, Christopher Rizzi, Akhil R. Shah, Peter A. Wilson, and Jerry M. Sollinger, *Lessons from the Army's Future Combat Systems Program*, Santa Monica, Calif.: RAND Corporation, MG-1206-A, 2012.

3 Sydney J. Freedberg, Jr., "Air Force ABMS: One Architecture to Rule Them All?" *Breaking Defense*, November 8, 2019.

4 James Lewis and Martin Fowler, "Microservices: A Definition of This New Architectural Term," blog post, martinfowler.com, March 24, 2014; and Amar Shan, "Heterogeneous Processing: A Strategy for Augmenting Moore's Law," *Linux Journal*, January 2, 2006.

5 Julian F. V. Vincent, Olga A. Bogatyreva, Nikolaj R. Bogatyrev, Adrian Bowyer, and Anja-Karina Pahl, "Biomimetics: Its Practice and Theory," *Journal of the Royal Society Interface*, Vol. 3, No. 9, 2006.

deception. None of the identified options observed every desired trait, but most observed a majority. We settled on the immune response to pathogens through agreement with the sponsor, in large part because the immune response contains a demonstration of both *heterogeneity* (the immune system is made up of different parts with unique capabilities) and *composability* (response actions are tailored and made up of differing parts, depending on the threat).

Immune Response Analysis Approach

As mentioned earlier, before settling on the immune response as a relevant case study, we surveyed the natural world for possible analogues to Mosaic warfare and found many useful examples, several of which focus on swarming behavior:[6]

- Many birds migrate biannually in flocks, assuming specific formations to save energy, boost efficiency, and minimize predation risk.[7]
- Fish derive many benefits from traveling in shoals, including defense against predators, optimized food gathering, and increased hydrodynamic efficiency.[8]
- Ants are commonly used as models for swarming behavior because of their foraging capabilities and construction of complex structures. Ants link their bodies together to form rafts to survive floods, assemble pulling chains to move food, and form bivouacs

---

6 Simon Garnier, Jacques Gautrais, and Guy Theraulaz, "The Biological Principles of Swarm Intelligence," *Swarm Intelligence*, Vol. 1, 2007.

7 C. J. Cutts and J. R. Speakman, "Energy Savings in Formation Flight of Pink-Footed Geese," *Journal of Experimental Biology*, Vol. 189, No. 1, 1994; and Anders Hedenström, "Adaptations to Migration in Birds: Behavioural Strategies, Morphology and Scaling Effects," *Philosophical Transactions of the Royal Society B*, Vol. 363, No. 1490, January 2008.

8 Jean-Guy J. Godin, "Antipredator Function of Shoaling in Teleost Fishes: A Selective Review," *Canadian Naturalist*, Vol. 113, No. 3, January 1986; D. J. Hoare, J. Krause, N. Peuhkuri, and Jean-Guy J. Godin, "Body Size and Shoaling in Fish," *Journal of Fish Biology*, Vol. 57, No. 6, 2000; and Tony J. Pitcher, "Functions of Shoaling Behaviour in Teleosts," in Tony J. Pitcher, ed., *Behaviour of Teleost Fishes*, Boston, Mass.: Springer, 1986.

and towers, as well as bridges and ladders to traverse rough terrain.[9]

- Honeybee swarms form large tree-hanging clusters made solely of bees attached to one another; these clusters respond to rain and wind by tuning the density of bodies and the cluster's surface area–to-volume ratio to maintain a near-constant core temperature, despite large fluctuations in ambient temperature.[10]

Although these biological systems show many similarities to Mosaic warfare, they are also rather predictable analogies to use and, critically, they all contain *homogeneous* swarms; that is, the elements participating in the swarm are of a uniform type. The human immune system, on the other hand—specifically, its response to foreign pathogens—uses many different elements with unique behaviors, exhibiting a *heterogeneity* that is desired of Mosaic warfare. Also, it has (to our knowledge) been largely ignored as an analogue for Mosaic warfare, despite being one that has been studied in similar contexts.[11] Table 2.2 lists our assessment of how the immune system compares with the Mosaic warfare vision along the six relevant metrics.

In this case study, we describe the immune response itself and draw out the comparison between this system and the Mosaic warfare vision. In particular, we detail five high-level analogies: (1) a *kill chain* used to detect, maintain custody of, and engage targets in a dynamic scenario; (2) *base defense*, wherein various resources and methods are used to effectively repel a variety of attacks on an installation; (3) *stra-*

---

[9] Nathan J. Mlot, Craig A. Tovey, and David L. Hu, "Fire Ants Self-Assemble into Waterproof Rafts to Survive Floods," *Proceedings of the National Academy of Sciences of the United States of America*, Vol. 108, No. 19, 2011; and Sulisay Phonekeo, Nathan Mlot, Daria Monaenkova, David L. Hu, and Craig Tovey, "Fire Ants Perpetually Rebuild Sinking Towers," *Royal Society Open Science*, Vol. 4, No. 7, July 2017.

[10] Samuel A. Ocko and L. Mahadevan, "Collective Thermoregulation in Bee Clusters," *Journal of the Royal Society Interface*, Vol. 11, No. 91, 2014; and Thomas D. Seeley and P. Kirk Visscher, "Group Decision Making in Nest-Site Selection by Honey Bees," *Apidologie*, Vol. 35, No. 2, March–April 2004.

[11] The immune response has been studied before to gain engineering insights. For example, see Stephanie Forrest, Steven A. Hofmeyr, and Anil Somayaji, "Computer Immunology," *Communications of the ACM*, Vol. 40, No. 10, October 1997.

Table 2.2
Comparison of the Immune System and Mosaic Warfare

Dimension	Immune System	Mosaic Warfare Goals
Fractionation	Multiple cell types communicate using chemical markers carried by the bloodstream or attached to the surface of pathogens and infected cells	Many platforms and platform types linked together in common data networks
Heterogeneity	Diverse set of cells with unique functions (e.g., each T cell responds to one specific epitope, and a given epitope is specific to a given pathogen, thereby necessitating the presence of many different types of T cells to respond to multiple pathogens)	Diverse number of multiservice/multiagency platforms included as needed
Rapid composability	White blood cells are recruited and signaled to reproduce based on presence of chemical markers	Rapid process for introducing new effects or capabilities, certification unclear
System architecture	Extremely complex and constantly evolving	Complex and rapidly evolving multidimensional relationships
Scalability and multiagent collaboration	Response actions automatically scale from localized attacks to large-scale symptoms, including fever and systemic infections	Joint multidomain assets that interoperate and scale easily up and down as necessary
AI/ML and autonomy	Completely autonomous (not directly controlled by the central nervous system)	Significant reliance on AI/ML and automation likely required

tegic preparations (intelligence gathering) for conflict; (4) *rapid mobilization* of forces in preparation for hostilities; and (5) the various risks and failures of military operations (such as fratricide). We sought not only to reveal the value of Mosaic warfare but also to uncover potential challenges facing such systems, allowing senior leaders to better prepare for and navigate future limitations of this type of approach.

The discussions on the immune system in Chapter Three are kept very focused, providing just enough detail to explore similarities to the Mosaic warfare vision. A detailed description of the variety and mechanisms of the immune system's behavior and response to pathogens could fill several textbooks; therefore, the analysis presented in our report will necessarily exclude many important functional details. Nevertheless, our analysis should suffice for drawing broad comparisons in approach and characteristics between the immune response and the Mosaic warfare vision.

Naval Integrated Fire Control—Counter Air Analysis Approach

We present a case study of the U.S. NIFC-CA project, draw lessons learned from the development of NIFC-CA, and apply these lessons, where possible, to the Mosaic warfare concept.[12] The focus of the NIFC-CA case study is to analyze the organizational construct, the experimentation process, and the documentation used to support NIFC-CA decisionmaking and budgets.

Table 2.3 compares NIFC-CA and the Mosaic warfare construct along our six dimensions. Despite the apparent differences between the two concepts, it is possible to draw important context and insights from the development of NIFC-CA.

This case study relies on three critical sources of information:

1. We performed a comprehensive literature review of unclassified documents on NIFC-CA, including documents by university-affiliated research centers (UARCs), federally funded R&D centers, and Naval Warfare Centers; Department of the Navy budget estimates; and other news articles relevant to NIFA-CA.

---

[12] We use NIFC-CA as a relevant case study because of its similarity to the Mosaic warfare vision. Indeed, this analogy also highlights that many of the Mosaic warfare concepts are not new. In fact, the original ideals of net-centric warfare align closely with Mosaic warfare's goals regarding ubiquitous communications, coordination, and adaptive kill chains. (Clay Wilson, *Network Centric Operations: Background and Oversight Issues for Congress, CRS Report to Congress*, Washington, D.C.: Congressional Research Service, March 15, 2007.)

Table 2.3
Comparison of Naval Integrated Fire Control—Counter-Air and Mosaic Warfare

Dimension	NIFC-CA Realities	Mosaic Warfare Goals
Fractionation	Limited number of platforms to be linked in the NIFC-CA data network	Many platforms and platform types linked together in common data networks
Heterogeneity	Platforms are a subset of those that make up a carrier strike group (CSG)	Diverse number of multiservice/multiagency platforms included as needed
Rapid composability	Lengthy process to test and certify any new effect or capability to the NIFC-CA data network	Rapid process for introducing new effects or capabilities, certification unclear
System architecture	Focused on naval aviation, simple relationships, statically defined	Complex and rapidly evolving multidimensional relationships
Scalability and multiagent collaboration	Currently focused on AEGIS-enabled ships, the SM-6, and naval aviation in support of CSG missions; no interoperability with the U.S. Air Force	Joint multidomain assets that interoperate and scale easily up and down as necessary
AI/ML and autonomy	Rule-based "first wave" artificial intelligence (AI)[a]	Significant reliance on AI/ML and automation likely required

NOTES: SM = Standard Missile.

[a] For DARPA's perspective on the "three waves of AI," see DARPA, "DARPA Perspective on AI," webpage, undated.

2. We held discussions with U.S. Navy subject-matter experts on the history, formation, and execution of NIFC-CA and any lessons learned and best practices they observed.
3. We used firsthand experience and direct knowledge from one of the authors, who served as the Naval Sea Systems Command's executive director, chief engineer, and chief technology officer, and from the Deputy Chief of Naval Operations for Warfare Systems, who is the primary resource sponsor for NIFC-CA.

The data and information gathered from these efforts were synthesized into a set of lessons learned that should apply to DARPA's Mosaic warfare project.

CHAPTER THREE

Lessons from the Immune System

Human bodies are essentially highly functional police states—the immune system is constantly monitoring the body's internal environment, ensuring that its molecular citizens are carrying out their respective roles. It learns and memorizes what is normal so that it can recognize when something unexpected happens. When an injury reaches a certain level of severity, the immune system activates cascades of molecular signals that recruit specialized immune cells to the site of injury, working to mount a defense and subsequently eliminate the threat. Invading pathogens, such as those that cause the common cold, chicken pox, and severe acute respiratory syndrome, are known to hijack cells and cellular communication systems, often reprogramming cellular machinery to commandeer the environment for themselves.[1] With just a few changes in terminology, this discussion could easily become about warfare. In this chapter, we will analyze how the immune system responding to pathogens has analogues with Mosaic

[1] Vorrapon Chaikeeratisak, Katrina Nguyen, Kanika Khanna, Axel F. Brilot, Marcella L. Erb, Joanna K. C. Coker, Anastasia Vavilina, Gerald L. Newton, Robert Buschauer, Kit Pogliano, Elizabeth Villa, David A. Agard, and Joe Pogliano, "Assembly of a Nucleus-Like Structure During Viral Replication in Bacteria," *Science*, Vol. 355, No. 6321, January 2017; Hua Niu, Qingming Xiong, Akitsugu Yamamoto, Mitsuko Hayashi-Nishino, and Yasuko Rikihisa, "Autophagosomes Induced by a Bacterial Beclin 1 Binding Protein Facilitate Obligatory Intracellular Infection," *Proceedings of the National Academy of Sciences of the United States of America*, Vol. 109, No. 51, December 18, 2012; and Natalie C. Silmon de Monerri and Kami Kim, "Pathogens Hijack the Epigenome: A New Twist on Host-Pathogen Interactions," *American Journal of Pathology*, Vol. 184, No. 4, April 2014.

warfare, ultimately arriving at several observations regarding the benefits, potential pitfalls, and desired traits for Mosaic warfare.

The Immune System

The immune system uses a complex array of protective mechanisms to control and eliminate pathogens. In short, the immune system protects the body from possibly harmful substances by recognizing antigens—or proteins—on the surface of the cell that are presented by invaders and destroying those invaders. This system has evolved to protect the host from a wide variety of pathogens, including bacteria, viruses, fungi, and parasites, that are themselves constantly evolving.[2] We will first describe a few of the elements of the immune system and will then discuss how it behaves in response to pathogens, touching on several interesting traits for which we will later draw analogues to Mosaic warfare.

Elements of the Immune System

This chapter is intended for an audience that has general familiarity with biology and the immune system. However, discussions will likely include several elements of the immune system with which the average reader is unfamiliar. To facilitate this discussion, in Table 3.1, we present a review of relevant elements of the immune system, including various cells, membranes, and chemical processes. This table is not meant to provide an exhaustive list of the components that make up the immune system.

Response to Pathogens

Figure 3.1 shows a simplified illustration of the immune system's response to pathogens, broken into two stages called *innate immunity* and *adaptive immunity*.

---

[2] David D. Chaplin, "Overview of the Immune Response," *Journal of Allergy and Clinical Immunology*, Vol. 125, No. 2, Supp. 2, February 2010.

Table 3.1
Selected Immune System Elements

Element	Purpose
Epithelial tissue	Tissue (e.g., skin) that covers surfaces and lines cavities within the body; primary functions include protection and filtration
Mucous membrane	Tissue that lines the digestive, respiratory, and reproductive tracts; stops particulates from entering the body
Epithelial cilium	Hairlike structure on the outside of a cell that lines the respiratory tract; this helps clean the mucous membrane and removes foreign particles from the body
Myeloid cell	Subtype of white blood cell made in bone marrow that engulfs and destroys pathogens, secretes cytokines, and triggers acute inflammatory and allergic responses; subtypes include granulocytes, neutrophils, monocytes, and macrophages
Cytokine	Small protein that guides the migration of cells in both the innate and adaptive immune systems (e.g., chemokines)
Lymphocyte	Subtype of white blood cell that includes B cells, T cells, and natural killer (NK) cells; recognizes antigens, produces antibodies, and destroys pathogens
Dendritic cell	Antigen-presenting cell that is responsible for initiation of the adaptive immune response
Phagocyte	Subtype of myeloid cell that helps protect the body by ingesting and destroying foreign particles, pathogens, and cell debris; subtypes include neutrophils, macrophages, monocytes, and dendritic cells
Antigen	Substance that evokes an immune system response, particularly the production of antibodies; importantly, antigens can also be structures that are part of the self, meaning that antigens are not always foreign
Antibody	Protein secreted by B cells in response to exposure to antigens that binds to the antigen and subsequently neutralizes pathogens
Memory cell	Type of lymphocyte that is capable of rapidly identifying and responding to a particular pathogen if it reappears

SOURCE: E. John Wherry and David Masopust, "Adaptive Immunity: Neutralizing, Eliminating, and Remembering for the Next Time," in Michael G. Katze, Marcus J. Korth, G. Lynn Law, and Neal Nathanson, eds., *Viral Pathogenesis*, 3rd ed., London, UK: Elsevier, Academic Press, 2016.

Figure 3.1
Simplified Model of Immune Response

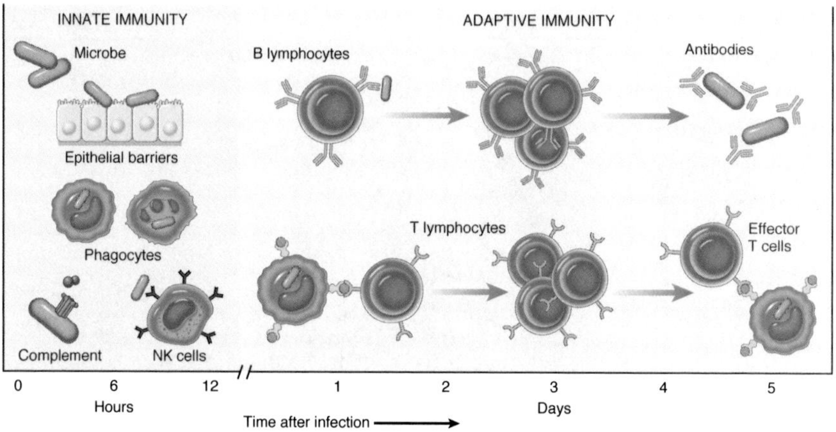

SOURCE: Vinay Kumar, Abul K. Abbas, and Nelson Fausto, eds., *Robbins and Cotran Pathologic Basis of Disease*, 7th ed., Philadelphia, Pa.: Saunders, 2004. Used with permission.

The nonspecific innate immune response provides the first line of defense against invading pathogens. Several components constitute the innate immune system, the most significant being physical barriers, including epithelial tissue (e.g., skin); the mucous membrane present in the respiratory, gastrointestinal, and genitourinary tracts; and *epithelial cilia* (hairlike structures on the outsides of cells) that clean the mucous layer after becoming contaminated with inhaled or ingested particles.[3] Myeloid cells, a family of blood cells manufactured in bone marrow, also play a significant role in innate immunity. Upon pathogen invasion, myeloid cells are recruited to local tissues via various chemokine receptors. Several subtypes of myeloid cells engulf pathogenic microbes and destroy them, while other subtypes participate in acute inflammatory and allergic responses.[4] Other examples of local acute immune responses include abscesses and granulomata, which are defensive

3 Chaplin, 2010.

4 Hiroshi Kawamoto and Nagahiro Minato, "Myeloid Cells," *International Journal of Biochemistry & Cell Biology*, Vol. 36, No. 8, August 2004.

mechanisms that trigger the body to sequester foreign invaders, such as bacteria or fungi, to keep them from spreading. This approach is inevitable in innate responses that lack specificity. Constantly renewed to reinforce strength, the innate immune response successfully wards off most incursions. However, the innate system cannot always recognize or eliminate infectious organisms, at which point the second line of defense, the adaptive immune system, enters the battlefield.[5]

The adaptive immune system manifests exquisite specificity for its target antigens, with responses based primarily on the antigen-specific receptors expressed on the surfaces of cells.[6] It should be noted that several components, such as NK cells, play roles in both innate and adaptive immune responses and readily blur the line between them.[7] So, although some aspects of the immune response detailed in this section might be considered parts of both the innate and the adaptive systems, we will focus our discussion on what is traditionally considered to be the adaptive immune response.

An intact adaptive immune response includes contributions from large subpopulations of lymphocytes, each population with a unique morphology and playing a distinct role. Lymphocytes have diverse antigen receptors for use in recognizing and repelling pathogen invaders. Mature lymphocytes have three main lineages—B cells, T cells, and NK cells—with B cells and T cells being the primary components of the adaptive immune response.[8] Once an invader has been identified, these cells generate specific responses that are tailored to eliminate specific pathogens and pathogen-infected cells. B cells are primarily responsible for *antibody-driven* immunity, producing antibodies that travel throughout the body in the bloodstream (conferring so-called humoral immunity), bind to antigens, and subsequently neutralize

---

5 Charles A. Janeway, Jr., Paul Travers, Mark Walport, and Mark J. Shlomchik, *Immunobiology*, 5th ed., New York: Garland Science, 2001.

6 Chaplin, 2010.

7 Eric Vivier, David H. Raulet, Alessandro Moretta, Michael A. Caligiuri, Laurence Zitvogel, Lewis L. Lanier, Wayne M. Yokoyama, and Sophie Ugolini, "Innate or Adaptive Immunity? The Example of Natural Killer Cells," *Science*, Vol. 331, No. 6013, January 2011.

8 Kawamoto and Minato, 2004.

them.[9] In contrast, cytotoxic T cells are best adapted to *cell-mediated* immunity, attacking the antigen and destroying cells that have already been infected.[10] Each T cell bears receptors to a single specific antigen, so a repertoire of T cells that can protect against a multitude of pathogens must include a very large number of cells. NK cells are cytotoxic and frequently serve to contain viral infections while the adaptive immune response is generating antigen-specific T cells that can clear the infection.[11]

Immunological memory, which is simply information storage, is a hallmark of adaptive immunity and is characterized by the long-term persistence of memory cells. Memory cells are subtypes of B and T cells formed following B- and T-cell activation by a specific pathogen, providing a lasting memory of that pathogen. These cells are able to recognize antigens previously encountered such that subsequent exposure to the same pathogen presents enhanced effector functions, mobilizing a stronger, faster response. This mechanism is used by vaccines, as we will discuss later.[12]

Communication within the immune system and between various components is mediated via cytokines and chemokines and the cell-surface receptors to which they bind. The elements operate by interacting with or attracting cells bearing specific receptors for them.[13] These various signaling mechanisms work together to activate and coordinate the many functions of the immune system, including up- and down-regulation of the immune response, lymphocyte homing, antibody

---

[9] William Hoffman, Fadi G. Lakkis, and Geetha Chalasani, "B Cells, Antibodies, and More," *Clinical Journal of the American Society of Nephrology*, Vol. 11, No. 1, January 2016.

[10] Mads Hald Andersen, David Schrama, Per thor Straten, and Jürgen C. Becker, "Cytotoxic T Cells," *Journal of Investigative Dermatology*, Vol. 126, No. 1, January 2006.

[11] Vivier et al., 2011.

[12] Hoffman, Lakkis, and Chalasani, 2016; and Ellen S. Vitetta, Michael T. Berton, Christa Burger, Michael Kepron, William T. Lee, and Xiao-Ming Yin, "Memory B and T Cells," *Annual Review of Immunology*, Vol. 9, 1991.

[13] In Janeway et al., 2001, see the section titled "Principles of Innate and Adaptive Immunity."

secretion, inflammation initiation, and targeting followed by subsequent elimination of pathogens.[14]

Activity Levels at Baseline and During Infection

Although the description thus far has focused on the adaptive immune system's response to invasions by serious pathogens, the reality is that such invasions are rare and do not represent the body's steady-state equilibrium, also known as *homeostasis*. Exposure to potential pathogens occurs on a spectrum, so the immune system must operate on a similar spectrum. Under usual circumstances in a healthy individual, the immune system operates at low levels compared with the activity that has been described thus far. The system is constantly responding to low-level attacks and working to maintain homeostasis, which is analogous to having small-scale conflicts around the globe that require the military's attention and resources but never escalate to large-scale combat operations (in large part because of the fact that the military has dedicated attention and resources to the mission). However, when the small-scale pathogenic attacks begin to overwhelm the body's state of equilibrium, the immune system rapidly ramps up its response, which manifests as physical symptoms of illness, including fatigue, fever, and inflammation.

T cells replicate at their maximum rate until they are no longer able to find available antigens. This control mechanism keeps the immune system from becoming overactive—a potentially fatal failure, as discussed in the next section. The level of immune system cell (specifically, T-cell) expansion also is related to the quantity of infectious agents and the affinity of the cells for those agents.[15] The greater the quantity of infectious agents and the higher the affinity of the T cells for the antigens, the larger the final number of T cells needed.

---

[14] Jianming Xie, Cristina M. Tato, and Mark M. Davis, "How the Immune System Talks to Itself: The Varied Role of Synapses," *Immunological Reviews*, Vol. 251, No. 1, January 2013.

[15] Andreas Mayer, Yaojun Zhang, Alan S. Perelson, and Ned S. Wingreen, "Regulation of T Cell Expansion by Antigen Presentation Dynamics," *Proceedings of the National Academy of Sciences of the United States of America*, Vol. 116, No. 13, March 2019.

The number of white blood cells varies widely during homeostasis, but a range of between 5,000 and 13,000 cells per mm^3, representing roughly a 3:1 variation in cells transiting the blood, is considered normal.[16] In the case of infections, this concentration can rise to as many as 50,000 cells/mm^3, representing a roughly 12:1 variation between the lower end of homeostasis and the upper end of response to infections.[17] This range is remarkably similar to the variation in U.S. military personnel (as a percentage of the total population) between World War II (WWII) and 2006, which peaked at around 8.5 percent during WWII and has reached a nadir of around 0.6 percent today (14:1). We see a similar relationship for spending as a percentage of gross domestic product, which reached a peak of 35 percent during WWII and is at a low of 4.2 percent today (8.3:1).[18] This is coincidental but reinforces the idea that both the immune system and the U.S. military experience variation of roughly an order of magnitude between peak mobilization and steady state.

Failures of the Immune System

Individual components of the immune system can overreact, misreact, or underreact. Overreaction to a given foreign substance manifests as an allergy. Reaction to an antigen that is part of one's own body is referred to as autoimmunity. Finally, several factors can lead to an underactive or compromised immune response.

Allergies originate when the immune system reacts to a substance that is harmless to most individuals, releasing such chemicals as hista-

---

[16] Svend Juul, Joseph S. Pliskin, and Harvey V. Fineberg, "Variation and Information in White Blood Cell Differential Counts," *Medical Decision Making*, Vol. 4, No. 1, 1984.

[17] Neil Abramson and Becky Melton, "Leukocytosis: Basics of Clinical Assessment," *American Family Physician*, Vol. 62, No. 9, November 2000.

[18] Defense Manpower Data Center, "DoD Personnel, Workforce Reports & Publications," webpage, undated; U.S. Bureau of the Census, "Chapter D: Labor," *Historical Statistics of the United States: Colonial Times to 1970*, Part 1, Washington, D.C.: U.S. Department of Commerce, 1975; and U.S. Office of Management and Budget, "Historical Tables," webpage, undated, Table 14.5.

mine and developing antibodies against the substance.[19] Upon repeated or sustained exposure, the severity of the allergic reaction might increase. Immune responses can vary from mild, such as coughing or a rash, to severe or even life-threatening—i.e., anaphylactic shock.[20]

Autoimmune disorders are another broad category characterized by inappropriate immune system responses. These disorders stem from an inability to properly distinguish between self and nonself, otherwise known as a breakdown of self-tolerance.[21] Genetic mutations in immune cells cause them to lose their ability to differentiate between pathogens and an individual's own cells, subsequently attacking healthy tissue and generating symptoms of disease unnecessarily.[22] Common autoimmune conditions include rheumatoid arthritis, multiple sclerosis, and Celiac disease.[23]

The human immunodeficiency virus (HIV) is an example of a virus capable of damaging the immune system beyond repair. HIV interferes with the immune system's ability to fight off invading pathogens, leaving the body susceptible to other diseases and infections. This means that people living with HIV often have higher rates of autoimmune disorders because of a weakened immune system.[24] In cases in which HIV progresses to acquired immunodeficiency syndrome (AIDS), the immune system has been severely damaged, making it

---

[19] Rudolf Valenta, "The Future of Antigen-Specific Immunotherapy of Allergy," *Nature Reviews Immunology*, Vol. 2, June 2002.

[20] Johns Hopkins Medicine, "Allergies and the Immune System," webpage, undated.

[21] G. H. Ring and F. G. Lakkis, "Breakdown of Self-Tolerance and the Pathogenesis of Autoimmunity," *Seminars in Nephrology*, Vol. 19, No. 1, January 1999.

[22] Alexis L. Franks and Jill E. Slansky, "Multiple Associations Between a Broad Spectrum of Autoimmune Diseases, Chronic Inflammatory Diseases and Cancer," *Anticancer Research*, Vol. 32, No. 4, April 2012; and Nikita Raje and Chitra Dinakar, "Overview of Immunodeficiency Disorders," *Immunology and Allergy Clinics of North America*, Vol. 35, No. 4, November 2015.

[23] Franks and Slansky, 2012.

[24] Yung-Feng Yen, Pei-Hung Chuang, I-An Jen, Marcelo Chen, Yu-Ching Lan, Yen-Ling Liu, Yun Lee, Yen-Hsu Chen, and Yi-Ming Arthur Chen, "Incidence of Autoimmune Diseases in a Nationwide HIV/AIDS Patient Cohort in Taiwan, 2000–2012," *Annals of the Rheumatic Diseases*, Vol. 76, No. 4, April 2017.

increasingly likely that the individual will develop opportunistic infections or cancers. For those patients for whom AIDS is fatal, death is often caused by one of these infections or cancers that an otherwise healthy immune system would be able to fend off.[25]

Analogues to Mosaic Warfare

The Immune Response Cast as a Kill Chain

A *kill chain* is a systematic process to target and engage an adversary to create desired effects. The military kill-chain model used in the Air Force is known as *F2T2EA*, which includes the following phases:[26]

1. Find: Identify a target.
2. Fix: Make an accurate determination of the target's location.
3. Track: Monitor the target's movement. Keep track of the target until either a decision is made not to engage the target or the target is successfully engaged.
4. Target: Select an appropriate weapon or asset to use on the target to create desired effects. Apply C2 capabilities to assess the value of the target and the availability of appropriate weapons to engage it.
5. Engage: Apply the weapon to the target.
6. Assess: Evaluate effects of the attack.

In this section, we will demonstrate that an analogous approach is used by the immune system. At each stage, we will define a potential military analogue to the immune response behavior.

Find/Fix: The immune system first identifies invaders by their *antigens*, which are proteins on the surface of the invading cells. Every

---

[25] Peter Piot, Michael Bartos, Peter D. Ghys, Neff Walker, and Bernhard Schwartländer, "The Global Impact of HIV/AIDS," *Nature*, Vol. 410, April 2001.

[26] John A. Tirpak, "Find, Fix, Track, Target, Engage, Assess," *Air Force Magazine*, July 1, 2000; and U.S. Air Force, "Dynamic Targeting and the Tasking Process," Annex 3-60, Maxwell Air Force Base, Ala.: Curtis E. LeMay Center for Doctrine Development and Education, last updated March 15, 2019.

cell has its own specific antigens, and cells that originate in the body carry what are called *self-antigens*, which are unique to that individual. A variety of immune cells present in the blood and lymph identify an invading cell by the fact that it does not carry self-antigens.[27] In an Air Force context, this stage could be represented by a proliferated force of sensitive but nondiscriminatory units that are hunting for adversary forces and that can raise the alarm when they encounter adversary forces.

Track: Once a threat is identified, immune cells release cytokines to communicate with other cells and control the response to the threat. Cytokines signal to immune cells that a pathogen is present and needs to be destroyed.[28] NK cells are among the first to be recruited to the site, staging a general attack on both the invader and the cells already infected.[29] At the same time, dendritic cells start to process the nonself-antigen material present on the surface of the invading cells, transforming themselves into antigen-presenting cells (APCs). These APCs then present the antigen material on their cell surfaces to B and T cells, exposing the makeup of the invader cells so the B and T cells can successfully locate the invading pathogen.[30] In a military context, we might consider the natural killers to be unsophisticated attack units with a low-cost weapon that attempt to blunt adversary attacks and impose attrition, accompanied by specialized sensor units that can uniquely identify and discriminate adversary units and communicate this information back to friendly forces.

Target/Engage: Depending on the makeup of the pathogen and the specific threat it poses, different lymphocytes are recruited to assist

---

[27] Chaplin, 2010.

[28] Mark J. Cameron and David J. Kelvin, *Cytokines, Chemokines and Their Receptors*, Austin, Texas: Landes Bioscience, 2000–2013; and Stephanie A. Condotta and Martin J. Richer, "The Immune Battlefield: The Impact of Inflammatory Cytokines on CD8+ T-Cell Immunity," *PLOS Pathogens*, Vol. 13, No. 10, 2017.

[29] Vivier et al., 2011.

[30] Ann-Katrin Hopp, Anne Rupp, and Veronika Lukacs-Kornek, "Self-Antigen Presentation by Dendritic Cells in Autoimmunity," *Frontiers in Immunology*, Vol. 5, No. 55, February 2014.

in the attack. B and T cells must receive signals from APCs before they can both recognize invading cells *and* destroy them.[31] B cells work rapidly to produce antibodies, which helps further identify and stop the invading pathogens.[32] T cells are activated in the presence of viruses and other pathogens that take over host cells. Fully activated T cells subsequently multiply to develop an army of T cells equipped with necessary weapons to defeat the threat.[33] This stage is similar to the assembly and deployment of customized response units or teams, which are formed to optimally defeat the specific adversary units detected.

Assess: Once the threat is gone, regulatory T cells work to slow and shut down the immune response.[34] Memory cells, which are subtypes of B and T cells, are able to recognize these same invading cells if and when they invade in the future, allowing the immune system to mount a faster, more effective response.[35] In a military context, this functionality would be performed by C2 units, in coordination with specialized sensor units, that de-escalate U.S. response forces and maintain a small force of staged response units in case the encountered units are observed again nearby.

Just as with F2T2EA, a broken link within this process affects the entire chain and desired outcomes. This analysis presents a potential mosaiclike power projection force loosely based on the F2T2EA kill chain.

From this comparison, we see that an immune system analogue for the Air Force targeting cycle might include (a) proliferated non-specific sensors looking for any sign of adversaries, (b) a second wave

---

31 See the section titled "Helper T Cells and Lymphocyte Activation" in Bruce Alberts, Alexander Johnson, Julian Lewis, Martin Raff, Keith Roberts, and Peter Walter, *Molecular Biology of the Cell*, 4th ed., New York: Garland Science, 2002.

32 Hoffman, Lakkis, and Chalasani, 2016.

33 Yu Tai, Qingtong Wang, Heinrich Korner, Lingling Zhang, and Wei Wei, "Molecular Mechanisms of T Cells Activation by Dendritic Cells in Autoimmune Diseases," *Frontiers in Pharmacology*, Vol. 9, No. 642, June 2018.

34 See the section titled "Principles of Innate and Adaptive Immunity" in Janeway et al., 2001.

35 Vitetta et al., 1991.

of general-purpose attack units and higher-fidelity sensors capable of threat identification and tracking, and (c) a third wave of specialized attackers well suited to the targets encountered. This type of approach seems best suited to situations in which the adversary's force posture is unknown and there is variation in how to attack each adversary unit (e.g., long-range weapons against units with surface-to-air missile support, cluster munitions against mechanized infantry).

The Immune Response Cast as Base Defense

In many respects, the human immune response is a perfect analogy for base defense. In both situations, we have a physically defended area with numerical superiority and defended assets surrounded by passive barriers and patrolled by active defensive systems. Furthermore, analogies can be drawn between the general immune response and many of the common defensive postures and systems, and the specific immune response can be cast as a mosaic-style defensive strategy that includes not only a resilient response but also the identification of threat system vulnerabilities and fabrication of new countermeasures.

General Defense

Joint Publication (JP) 3-10.1 explains the tactics and techniques to be employed by joint forces in defense of their bases.[36] Key stages of base-defense activity are Detect, Warn, Deny, Destroy, and Delay. Incursions must be detected, the base must be warned, attackers must be denied entry to the base, the attacking force must be destroyed (if possible), and it must be delayed (if destruction is not possible). We see many of these behaviors in the description of general immune response.

The base perimeter (and fencing, or other perimeter boundary) is analogous in many ways to the outer skin of a human being, but there is a key difference in the base-defense analogy. Although the immune response is typically not activated until a pathogen breaches the skin and enters the interior of the body, base defense begins with surveil-

---

[36] JP 3-10.1, *Joint Tactics, Techniques, and Procedures for Base Defense*, Washington, D.C.: Chairman of the Joint Chiefs of Staff, July 23, 1996.

lance of the area around a base to detect and identify threats before they attempt to breach the perimeter.

The behavior of myeloid cells—which quickly amass near the infection site, engulf and destroy the invaders, and alter the environment to delay the spread of infection—is analogous to base-defense forces that are quickly alerted and assembled to repel or delay invasion. They are general-response soldiers, armed with standard weaponry to repel invasion.

The case of an abscess, where the immune response engulfs an infection to prevent its spread, is an interesting case for study in this analogy. One can envision a scenario in which the threat is not readily identified, such as one in which attackers are disguised as friendly personnel.[37] The immune response suggests one way in which a Mosaic warfare system might handle such a threat if there is evidence of hostile activity but no positive identification of the threat: by cordoning off the area and sacrificing those friendly systems within to prevent further spread of the hostile forces, at least until sufficient capability is brought to bear to identify and suppress the threat. Furthermore, if such a response were likely to be necessary, it might inform development of the rules of engagement for Mosaic warfare to ensure that such a response occurs outside the perimeter of the base to minimize collateral damage.

Adaptive Defense

In addition to a rapid response from base forces, JP 3-10.1 calls for the employment of response forces to attack the invading force. These response forces should consist of specialized units, such as anti-armor or indirect fires weapons, as needed and available to repel the attacking force. This is analogous to the arrival of antigen-specific lymphocytes (B cells, T cells, and NK cells) that can recognize and respond to the antigens present in the infection.

---

[37] Consider, for example, recent base attacks in which insurgent groups infiltrated forces being trained at a base to gain access and execute an attack from within; see Patricia Mazzei, Thomas Gibbons-Neff, and Christine Hauser, "Trainee on Military Base Mounts Deadly Attack," *New York Times*, December 6, 2019.

The Mosaic warfare construct is observed in the response of lymphocytes to new antigens. These cells are able to adapt to the presence of new antigens, generate antibodies and effector lymphocytes that can destroy the new antigens, and preserve these capabilities for a more rapid response if the antigens are encountered again. In a base-defense context, this is akin to rapid collection of intelligence against threats, the design of new countermeasures (such as a sensor that is tuned to detect the unique signature of a threat), and the rapid fabrication of those components and integration into defense forces (e.g., mounting that sensor onto an unmanned aerial vehicle and using those data to cue and guide existing weapon systems or countermeasures). This approach would depend on significant technological advancements, including automatic design and fabrication of capabilities in the field, but it would represent the implementation of a mosaic vision for base defense.

Vaccines Cast as Strategic Preparation

Intelligence and planning are critical pieces of strategic preparation for conflict, providing knowledge of an adversary's capabilities, plans, and intentions. The goal is that this knowledge is obtained well in advance of any action taken by the adversary, allowing forces sufficient time to mount an appropriate defense. To find an analogy in the biological realm, we look to vaccines. Memory cells, described earlier, circulate through the bloodstream until they recognize an antigen introduced during a prior infection or vaccination. When they encounter an antigen for a second time, these cells mount a rapid and strong immune response. Since the introduction of the smallpox vaccine in 1796, following the discovery of immune response development, society has used vaccines as an artificial means of stimulating this response. Vaccines trigger the formation of memory cells by exposing the body to a safer variant of the disease,[38] such that the risk of succumbing to the infection is minimal. By inoculating a patient with a vaccine, medical

[38] This can be a similar strain that is less dangerous or a damaged or dead sample of the pathogen.

professionals provide the patient's immune system with a chance to observe the threat and determine how to defeat it.

Vaccine development can employ several different strategies, including using a weakened virus, an inactivated virus, or only part of the virus. Each strategy has significant advantages and disadvantages; however, each necessitates a deep knowledge and understanding of the pathogen, including how it infects, how it reproduces, and how it interacts with the *host* (the individual infected) and the host's immune system. To create a vaccine, a researcher must know enough about the infection to be able to mimic it, because administering a vaccine essentially mimics introducing an infection on a very small scale. This foundational research is like the significant effort inherent in the development and deployment of intelligence-gathering assets, including human intelligence sources (spies) and technical intelligence sources (such as listening posts or surveillance satellites).

Once a patient is vaccinated, the patient's immune system reacts to the vaccine as if it were a pathogen and responds with the elements of innate and adaptive immunity discussed earlier in this chapter. Eventually, lymphocytes are generated that can detect and destroy the pathogen, and some of those persist as memory cells.

To find an analogous stage in the defense realm, we look to battlefield and tabletop exercises (or wargames), where an adversary force structure and capability are posited and various responses are tested. This process can result in an assessment of the gaps or shortfalls where improvement is required. At this point, the military can decide where to invest resources to close the identified gaps. A common solution involves the procurement of new technologies and changes to tactics, techniques, and procedures.

This process of running through an exercise, conducting a gap assessment, and deciding where to invest resources to close the gaps is both challenging and very time-consuming. The existence of a mosaic system presents a promising alternative. Because the control system algorithms within a mosaic architecture are likely to be automated, one could instantiate them within a simulation with high-fidelity physics models and allow the mosaic system to respond naturally to the new threat capability or behavior.[39] The system could be run repeatedly until a proper response action is devised, and the final state of the mosaic architecture could then be used to update the control algorithms for fielded systems. The success of such a capability is dependent on many technological improvements, such as neural networks that can devise new behaviors and distributed controls schemes and then learn from the outcomes of a simulated conflict. The consideration of new technologies or system upgrades could be as simple as programming their effects into the simulation environment.

Immune Response Cast as Rapid Mobilization

As we described at the beginning of this chapter, the steady state for the immune system, referred to earlier as *homeostasis*, is often perceived as dormancy but actually represents a near-constant state of infection and low-level response. This state is analogous to a military posture with distributed presence and the capacity to rapidly respond to threats in any theater or domain, and it is like how the U.S. force is currently postured, with forward bases around the world that are clustered near strategic territory and potential conflict zones.

When an infection persists, the level of activity changes. Local elements of the immune system begin to signal for assistance, triggering fever and inflammation responses in an attempt to slow the pathogen's progress. In a military context, this is easily compared to a call for reinforcements that is tailored to the hostile force (such as a close-air support attack to engage enemy troops), as well as steps taken to seize control of the battlefield (such as wide-scale jamming to suppress enemy communications).

Finally, in response to a large-scale infection, the immune system increases production of white blood cells, such as B and T cells. As discussed earlier in this chapter, the scale of this increase is remarkably

---

[39] This is the basic concept behind OpenAI Gym, a simulation environment in which AI agents can be exposed to repeated and varied trials. It is popular with a branch of machine learning called *reinforcement learning*, in which agents respond to rewards (both positive and negative) based on simulation outcomes. For an overview of the concept, see Dave Gershgorn, "Elon Musk's Artificial Intelligence Group Opens a 'Gym' to Train A.I.," *Popular Science*, April 27, 2016.

similar to the difference in size and budget for the U.S. military from peacetime to full-scale war.

However, key differences between the military and the immune system are scale and distribution. The bloodstream represents a very rapid means of transporting newly produced cells to the site of an infection. In the military context, build-up and draw-down cycle is reflected in a time-phased force deployment data plan,[40] which outlines which units arrive in theater as a function of days before or after initiation of hostilities and can span many months before full mobilization is complete. It is becoming increasingly clear that full mobilization would take too long, and many conflicts could be lost before the full contingent of forces arrive in theater. For example, recent wargaming of a Baltic conflict with Russia predicts that the conflict would be won or lost within the first 60 hours.[41]

The challenges that come with the amount of time required for a full mobilization could be addressed by distributed production facilities, provided that the ability to ramp up production quickly is sufficient. This idea is particularly promising if the distributed production facilities are able to respond in real time to demand signals on the battlefield. If those developing Mosaic warfare are to learn from the immune system, one key trait might be the ability to rapidly produce the necessary systems when and where they are needed.

Failures of the Immune Response Cast as Risks to Distributed Operations

Although it is possible to learn much about the traits that a Mosaic warfare system might have or the benefits it might exhibit from a comparison with the immune system's pathogen response, it is also important to glean information about risks and shortcomings. There are various ways in which the immune system can fail, ranging from a lack of

---

[40] JP 3-35, *Deployment and Redeployment Operations*, Washington, D.C.: Chairman of the Joint Chiefs of Staff, January 10, 2018.

[41] David A. Shlapak and Michael W. Johnson, *Reinforcing Deterrence on NATO's Eastern Flank: Wargaming the Defense of the Baltics*, Santa Monica, Calif.: RAND Corporation, RR-1253-A, 2016.

response to threats to inadequate or untimely responses to overactive responses that perceive a threat where none exists.

In a military context, these deficiencies remind us of the dangers of a highly adaptive and dynamic system, such as Mosaic warfare. It will be difficult to know ahead of time how this novel approach will react to certain threats and environments. Through analysis of the ways in which the immune system fails, we can articulate some likely failure mechanisms for which DARPA should prepare.

A cytokine storm occurs when too many lymphocytes are attracted to an area, resulting in damage to the nearby tissue. An analogous military situation might result from a breakdown in communications, errors in inference that overestimate the size of the threat, or adversary attempts to inject false messages into U.S. C2 systems to redirect U.S. forces. There is no one way to guarantee that a fielded Mosaic warfare system could avoid this challenge, but it underscores the need for robust, reliable, and secure communications among the component systems.

Autoimmune disorders arise when immune cells lose the ability to distinguish between host cells and pathogens. In a military context, Identification Friend-or-Foe transponders are one of many ways in which adversaries are discriminated from friendly forces. Nevertheless, these systems are not perfect, and *fratricide* (the unintentional killing of friendly forces) persists in conflict. This phenomenon serves to remind us that the problem of fratricide will persist in the context of Mosaic warfare and could be devastating if, for example, an entire class of platforms were systematically misidentified as adversaries.

Immune deficiencies arise when a person's immune system fails to function (as occurs in individuals with HIV), in which case the virus attacks white blood cells, destroying their ability to respond to other pathogens. In a military context, this example serves as a reminder that the component systems that make up Mosaic warfare will themselves be subject to attack from both kinetic and nonkinetic threats. Building redundancy into critical elements in the network will help ensure that an attack that systematically suppresses a given type of component will not hamper the network's ability to function. In other words, there should be multiple ways to sense every threat system, multiple ways to track and target them, and multiple ways to engage. Thus, if an adversary fields a targeted countermeasure (such as a defense against radio frequency–guided missiles), the system can deploy alternatives (such as infrared-guided or command-guided missiles).

We have identified a biological system with mosaic attributes—the adaptive immune response—that is analogous to the Mosaic warfare vision. The immune system points to both potential benefits and potential limitations of the mosaic approach. This comparison allows us to draw some insights for how mosaic systems should be organized for warfare.

Potential Benefits of a Mosaic System

Mosaic systems are highly **adaptable.** Although it is possible to build a system that can adapt to predetermined stimuli or change within a narrow range of changing scenarios, a mosaic approach is preferred for systems that can adapt to new stimuli. With new stimuli comes the need to fundamentally change or eliminate parts of a system, so only a whole containing many functioning parts can survive while adapting to new environments.

Mosaic systems are also **resilient**, as this case study—and our previous experience—demonstrates. Mosaic systems incorporate an inherent amount of redundancy in that individual entities (cells or platforms) might not be critical to the survival of the system. Small failures are lost in the system, and larger failures are held in check; there is no single point of failure, because many agents can perform each task. Critically, this redundancy is not merely duplication but rather the existence of different mechanisms to achieve an effect, such that if one mechanism is neutralized by some adversary capability, another is still available to achieve the desired effect. These systems are *compensatory*, which means that that if one part is weak or nonfunctional, another part can typically step in. Although the substitute might or might not be as efficient, it allows for successful completion of the task.

Potential Limitations and Vulnerabilities of a Mosaic System

With the resiliency of a mosaic system also comes a level of **inefficiency**, which could be a limitation. The redundancies in the system can cause some inefficiencies and duplication of effort. For similar reasons, a mosaic system might be significantly more **difficult to control** than a *monolithic system* in that there is no central authority and the number of platforms to direct is likely to be much larger.[42] Although these limitations might be obvious, they are worth pointing out.

Traits Needed for a Successful Mosaic System

Mosaic warfare is often predicated on such traits as fractionation, heterogeneity, and composability. Our analysis of the immune response as an analogy for Mosaic warfare has highlighted several additional, nonobvious, and interesting traits that a successful Mosaic warfare system should exhibit. We note that the immune system analogy is inherently a defensive scenario in which numerical superiority and proximity of forces are advantages. Our comparisons, as drawn, are agnostic to these facts, but the underlying success of the immune system depends on the ability to muster large numbers of both general and specialized response forces and the ability to communicate information to reserve forces regarding the location and composition of adversary forces. This is an important caveat to the traits outlined in the paragraphs that follow.

First and, we believe, foremost is robust communication.[43] Several steps of the immune response rely on the transmission of information via chemical signals in the bloodstream. If this information is not transmitted to the B cells, T cells, NK cells, and other myriad components of the immune system, then it will be unable to amass the variety and quantity of response cells to successfully ward off pathogens. Similarly, if mosaic systems are unable to relay the initial detections, locations, and classifications of threats to the necessary special-

---

42 *Monolithic* describes a system or organization that is "large, powerful, indivisible, and slow to change." (Lexico.com, "Monolithic," webpage, undated.) Canonical examples include the F-35 multi-mission aircraft and the *Ford*-class aircraft carrier. Admittedly, this characterization is problematic in that the systems are not indivisible per se; but they are indivisible because they assign all capabilities together, while a mosaic arrangement is free to assign the capabilities separately.

43 By a *robust* communications system, we mean one that reliably relays information and is resistant to attempts to suppress communications and spoofing by an adversary.

ized response units, this approach will fail. Communication is critical to any modern military force, but dependence on reliable and robust communications is increased by Mosaic warfare's disaggregated nature.

Second, a rapid, tailored response to a wide array of possible threats requires the forward staging of a large force, improvements in rapid mobilization and deployment of reserve forces, or distributed rapid-production facilities. This is true of any force structure, but it is even more critical to Mosaic warfare given its reliance on large numbers of small platforms.

Third, the ability of the immune system to adapt quickly to previously unseen threats is critical to its continued success. The approach is more similar to an iterative experimental approach than to careful observation and analysis that characterize modern military systems acquisition. If Mosaic warfare is to be presented with unknown or previously unobserved adversary units and force structures, it might require the ability to quickly synthesize new "mosaics" (or collections of available "tiles") and assess their success, iterating and experimenting until a successful response is identified. This potential need contrasts with a more conventional approach, which would rely on collection of intelligence against the new threat throughout a system's acquisition process and reliance on R&D laboratories to develop countermeasures for acquisition. If successful, the mosaic response to novel threats could prove to be a critical capability if adversary acquisitions become opaque to U.S. intelligence collections.

Final Thoughts

In weighing the pros and cons of a given approach, we suggest that it is helpful to be able to look at the evolution of an analogous biological system—in this case, the immune response as an analogue for Mosaic warfare. If a given behavior or system does not evolve, this does not necessarily mean that the behavior or system is not beneficial. This specific behavior or system could be a remnant of what was once useful but no longer serves a purpose (e.g., vestigial organs). Similarly, if it *does* evolve, it is not necessarily the optimal way to solve a problem. However, if it does evolve, it is proof that the behavior or system was beneficial at some point during the evolution of the system.

Ample research has examined the evolution of the adaptive immune system and why it might have arisen. Although many hypotheses have been offered, and it is unknown exactly why or how it evolved, we do know that it has evolved over the course of nearly 500 million years to exhibit mosaiclike properties,[44] meaning that the mosaic characteristics have conferred some evolutionary advantage. From this, we infer that the distributed (mosaiclike) approach might have an advantage over an approach that focuses on a smaller number of more-capable agents when defending against and subsequently attacking and eliminating a threat.

---

[44] Martin F. Flajnik and Masanori Kasahara, "Origin and Evolution of the Adaptive Immune System: Genetic Events and Selective Pressures," *Nature Reviews Genetics*, Vol. 11, No. 1, 2010.

CHAPTER FOUR

Lessons from the Navy

The Naval Integrated Fire Control—Counter Air Project

Since 1996, the NIFC-CA project has been developing a family-of-systems (FoS) capability to defeat overland cruise missiles and other over-the-horizon (OTH) air warfare threats. A joint letter from former Under Secretary of Defense for Acquisition and Technology Paul Kaminski and former Vice Chairman of the Joint Chiefs of Staff William A. Owens, dated January 11, 1996,[1] provided direction to ensure that developmental systems could detect cruise missiles that could change course and speed and that were masked by the curvature of the earth, coastal mountains, hills, and other variations in terrain. This letter also initiated the overland cruise missile–defense (OCMD) SoS, which included the Army's aerostat program, the Navy's E-2D, and the Air Force's E-3 early warning aircraft programs and advanced interceptor-seeker development.[2]

Six years later, the OCMD SoS program was officially reorganized as the NIFC-CA project in a joint letter signed by the Assistant Secretary of the Navy for Research, Development and Acquisition and

[1] Undersecretary of Defense for Acquisition and Technology and Vice Chairman of the Joint Chiefs of Staff, "Land Attack Cruise Missile Defense (CMD)," joint memorandum, Washington, D.C., January 11, 1999.

[2] Jeffrey H. McConnell and Lorra L. Jordan, "Naval Integrated Fire Control—Counter Air Capability-Based System-of-Systems Engineering," *Leading Edge*, February 2013.

the Vice-Chief of Naval Operations, dated October 11, 2002.[3] The joint memo rescoped the project to defeat the OTH manned fighter and the OTH anti-ship cruise missile threats, in addition to the original OCMD mission. The joint memo also directed the Program Executive Officer for Integrated Warfare Systems (PEO IWS) to set up a NIFC-CA Systems Engineering and Integration Project Office to integrate the elemental programs to support the development and fielding of NIFC-CA. The NIFC-CA project was directed to execute a capabilities-based acquisition project by levying minimal requirements onto the component systems while deriving SoS capability from the federation of these independent systems. Eventually, the NIFC-CA project settled on an FoS engineering model that combined multiple sensors through Integrated Fire Control (IFC)–compliant combat systems to support extended-range active missiles.[4] The NICA-CA FoS includes three complete kill chains, as depicted in Table 4.1. Each FoS kill chain consists of elevated and surface sensors, a sensor network, a weapon control system, and an active missile.[5]

NIFC-CA did not receive a formal declaration of IOC, because it was not a formal acquisition program. It is considered to have achieved IOC in 2014, when the E-2D achieved IOC, thus completing the first NIFC-CA kill chain.[6] The USS *Theodore Roosevelt* CSG was the first NIFC-CA–capable CSG and later completed combat system ship qualification trials and integrated testing, in July 2015.[7]

---

3 Assistant Secretary of the Navy for Research, Development and Acquisition and Vice Chief of Naval Operations, "Updated Responsibilities for Management of Naval Integrated Fire Control—Counter Air (NIFC-CA)," joint memorandum, October 11, 2002.

4 McConnell and Jordan, 2013.

5 McConnell and Jordan, 2013.

6 Majumdar, 2014.

7 John F. Morton, "The Aegis Warship: Joint Force Linchpin for IAMD and Access Control," *Joint Forces Quarterly*, Vol. 8, No. 1, January 1, 2016.

Table 4.1
The Naval Integrated Fire Control—Counter-Air and Mosaic Warfare Family of Systems

System of Systems (Kill Chain)	Remote Sensors	Sensor Network	Weapon Control System	Active Missile
From-the-Air	E-2D F-18 E/F	Link-16	F-18 E/F	AMRAAM
From-the-Sea	E-2D JLENS	CEC	AEGIS ACB-12	SM-6
From-the-Land	E-2D JLENS AN/TPS-59 G/ATOR	CTN	CAC2S	TBD

SOURCE: McConnell and Jordan, 2013.

NOTE: ACB = Advanced Capability Build; AMRAAM = Advanced Medium Range Air-to-Air Missile (AIM-120D); CAC2S = Common Aviation Command and Control System; CEC = Cooperative Engagement Capability; CTN = Composite Tracking Network (CEC network hosted on U.S. Marine Corps land-mobile vehicles); G/ATOR = Ground/Air Task-Oriented Radar; JLENS = Joint Land Attack Cruise Missile Defense Elevated Netted Sensor System; TBD = to be determined.

Naval Integrated Fire Control—Counter Air Organizational Construct

Over time, the Navy solidified its approach to managing NIFC-CA into a complex organizational structure with oversight from PEO IWS-7, the NIFC-CA project office, and a collaborative Government/Industry Systems Engineering and Test Team, composed of personnel from the NIFC-CA project office, government Warfare Centers and laboratories, UARCs, and industry.

Figure 4.1 shows the original five pillar programs that formed the foundation of NIFC-CA, each of the program managers from these major acquisition programs was given an interface to PEO IWS 7D, the project manager for NIFC-CA, and provided staff to the Collaborative Government & Industry System Engineering & Test Team. In 2015, the Army suspended the JLENS program.[8] Each of the four

[8] JLENS was operationally suspended immediately following an incident in which one of the prototypes came untethered and dragged a tow cable across Maryland and Pennsylvania. The 2017 National Defense Authorization Act cut the JLENS program budget from

Figure 4.1
Naval Integrated Fire Control—Counter Air NIFC-CA System-of-Systems Management Structure

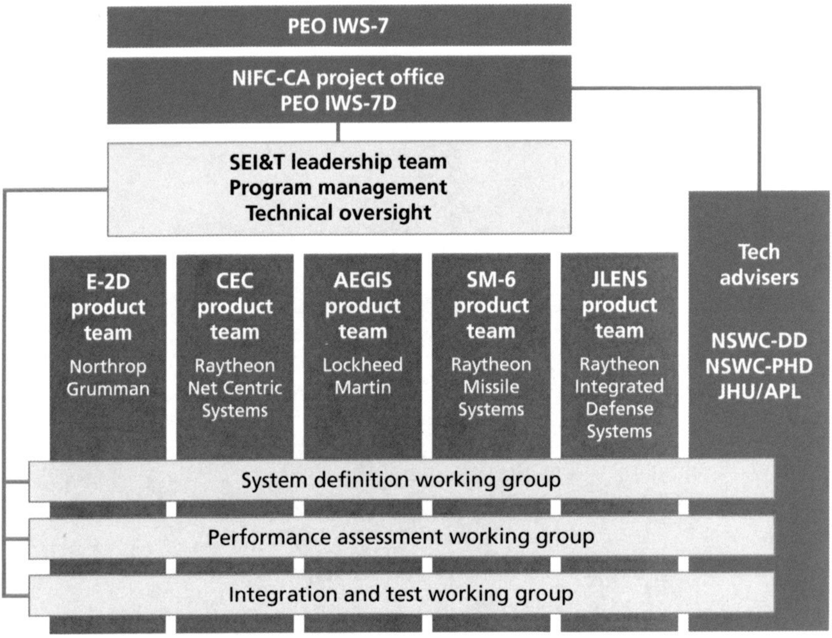

SOURCE: Adapted from McConnell and Jordan, 2013.
NOTE: JHU/APL = Johns Hopkins University/Applied Physics Laboratory; NSWC-DD = Naval Surface Warfare Center—Dahlgren Division; NSWC-PHD = Naval Surface Warfare Center—Port Hueneme Division; SEI&T = System Engineering, Integration and Test.

remaining pillar programs provides a critical element to the From-the-Sea (FTS) kill chain described in Table 4.1; these programs provide the necessary components to close an integrated fire control kill chain. Figure 4.2 provides a more detailed description of each of the original five pillar programs.

NIFC-CA is not an official acquisition program but rather a systems engineering and test project, and its capability is not derived from an initial set of requirements leading to component program selection.

$45 million to $2.5 million, ostensibly to cover costs necessary to close out the program (Jen Judson, "Congress Nails Runaway Blimp's Coffin Shut," Defense News, May 27, 2016).

Figure 4.2
Original Naval Integrated Fire Control—Counter Air Pillar Programs

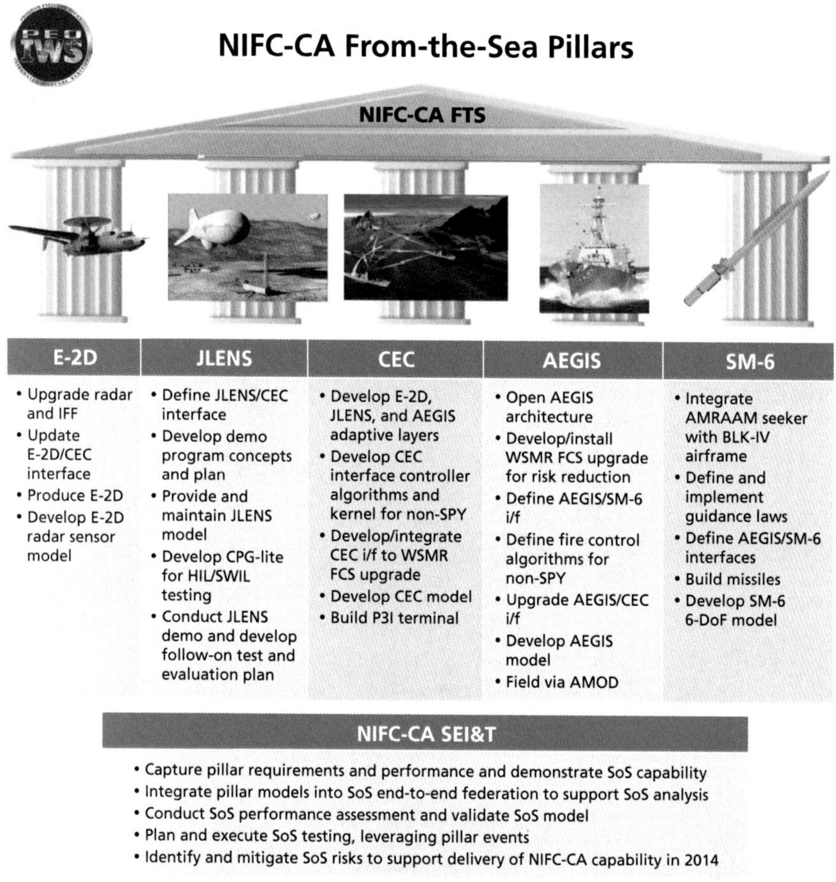

SOURCE: Adapted from Jim Syring, "Navy IAMD Capabilities," briefing slides for NDIA State of IAMD Symposium, Washington, D.C.: Program Executive Office, Integrated Warfare Systems, July 12, 2012.
NOTE: AMOD = AEGIS Modernization; AMRAAM = Advanced Medium Range Air-to-Air Missile; BLK-IV = Block IV; CPG = Communication and Processing Group; DoF = degrees of freedom; FCS = Future Combat Systems; HIL = hardware-in-the-loop; i/f = interface ; P3I = Pre-Planned Product Improvement; SWIL = software-in-the-loop; WSMR = White Sands Missile Range.

NIFC-CA capabilities are derived from FoS performance predictions by engineering analysis, by FoS modeling and simulations that predict the expected performance of the pillar programs, or both. Therefore, the only funding in the Navy budget that is specifically for NIFC-CA is research, development, test, and evaluation (RDT&E) funding for the integration and interoperability of the pillar programs and other programs, such as the Joint Strike Fighter (JSF) and the F/A-18 E/F strike fighter. NIFC-CA does not have dedicated procurement funding and does not procure any hardware. Any hardware that is procured for NIFC-CA capabilities is funded by the pillar programs' procurement funding.

As a result of this focus on system engineering rather than on procuring any additional hardware, NIFC-CA's budget allocation in fiscal year (FY) 2017–2020 was on the order of $35 million per year of RDT&E funding,[9] compared with hundreds of millions of dollars in total annual RDT&E funding and more than $1 billion in annual procurement funds for the remaining four pillar programs (AEGIS, E-2D, CEC, and SM-6).[10]

As mentioned earlier, JLENS, one of NIFC-CA's five pillar programs, was suspended in 2015. In a typical acquisition program, such a major event likely would have placed NIFC-CA in breach of some

---

[9] The President's budget request was $33.4 million for NIFC-CA in FY 2019, with an expected $35.4 million request in FY 2020, compared with roughly $25 million and $25.4 million in FY 2017 and FY 2018, respectively. (Department of the Navy, *Department of Defense Fiscal Year (FY) 2019 Budget Estimates: Research, Development, Test & Evaluation, Navy*, Justification Book Vol. 3, Budget Activity 5, Washington, D.C., February 2018c, pp. 631–640.)

[10] For example, the 2019 President's budget requested $884 million in procurement funds for four E-2D aircraft, $490.2 million in procurement funds for 125 SM-6, $130.5 million in RDT&E funds for CEC, $165.9 million in RDT&E funds for SM-6, and $223.5 million in RDT&E funds for E-2D aircraft. (Department of the Navy, 2018c, pp. 227–274; Department of the Navy, *Department of Defense Fiscal Year (FY) 2019 Budget Estimates: Aircraft Procurement, Navy*, Justification Book Vol. 1, Budget Activity 01–04, Washington, D.C., February 2018a, pp. 125–135; Department of the Navy, *Department of Defense Fiscal Year (FY) 2019 Budget Estimates: Research, Development, Test & Evaluation, Navy*, Justification Book Vol. 1, Budget Activities 1, 2, and 3, Washington, D.C., February 2018b; Department of the Navy, *Department of Defense Fiscal Year (FY) 2019 Budget Estimates: Research, Development, Test & Evaluation, Navy*, Vol. 5, Budget Activity 7, Washington, D.C., February 2018d, pp. 37–62; Department of the Navy, *Department of Defense Fiscal Year (FY) 2019 Budget Estimates: Weapons Procurement, Navy*, Vol. 1, Washington, D.C., February 2018e, pp. 89–100; and Department of the Navy, *Department of Defense Fiscal Year (FY) 2019 Budget Estimates: Aircraft Procurement, Navy*, Vol. 1, Washington, D.C., March 2019.)

of its requirements. Although this event might not have necessarily doomed a formal acquisition program, it certainly would have triggered a review of the requirements to determine whether a suitable change would meet the requirements or whether they could be waived. Because NIFC-CA lacks formal requirements and its budget came from RDT&E funding, no such formal review was required. However, it is critical to note that the functionality of JLENS was largely redundant with that of the E-2D pillar (both provided the detection and tracking data to start a NIFC-CA kill chain, albeit from different operating locations). Had the SM-6 or CEC pillar program been canceled, the outcome for NIFC-CA would have been far worse, and it is not clear that the project would have continued.

Lessons Learned

From this review of NIFC-CA's organizational construct, we draw an important lesson that can be applied to the previously mentioned prerequisite of maintaining current force structure and programs of record for migrating to the mosaic concept: The Navy's decision to structure NIFC-CA as a project with minimal (approximately $35 million per year) RDT&E funding for integration and interoperability of the pillar programs rather than to form a separate acquisition program with significant procurement funding resulted in a small increase in funding requirements on the underlying pillar programs of record. However, this decision allowed NIFC-CA to survive the turbulence of budget cycles and reviews, including the cancellation of one of its core pillars (JLENS).

Naval Integrated Fire Control—Counter Air Experimentation Approach

Experimentation efforts for NIFC-CA can be traced back to a January 1994 Army Mountain Top Experiment called "Mountain Top" to address the emerging threat of defending against land-based cruise missiles.[11] The Earth's curvature, and the low altitude of cruise missiles, prevents their detection at long range. When coupled with the ability of cruise missiles to maneuver in unpredictable paths, this presented a significant challenge for cruise missile defense. The Mountain Top Experiment attempted to integrate an AEGIS cruiser, a Patriot battery, and an experimental mountaintop radar, networked together with the Navy's Mk74 Fire Control System via the Navy's CEC, and successfully demonstrated the intercept of a cruise missile from ground-based assets.[12]

From the success of this experiment, the Navy began developing the CEC acquisition program to engage cruise missiles at greater distances from U.S. Navy cruisers and destroyers. CEC evolved into an Acquisition Category (ACAT) I acquisition program and has become one of the pillar acquisition programs for the NIFC-CA FTS kill chain. This experiment laid the foundation for future experiments and testing events for NIFC-CA projects.

Experiments or testing events to increase the capability of NIFC-CA follow a disciplined process of hypothesis, prediction, and measurement (HPM). For example, to increase the number of NIFC-CA cooperative engagement modes (CEMs), experiments were developed using the HPM methodology:

- **Hypothesis:** NIFC-CA will allow the expansion of the fleet tactical grid out to the maximum kinematic range of weapons by using architectures through expanding networks to increase sensor and fire control data.
- **Prediction:** There will be increased sensors contribution and spectral maneuver, distributed engagement options, and expanded battlespace awareness.

---

[11] Sharon Watkins Lang, "SMDC History: Army Mountain Top Experiment," webpage, U.S. Army, January 18, 2018.

[12] Lee O. Upton and Lewis A. Thurman, "Radars for the Detection and Tracking of Cruise Missiles," *Lincoln Laboratory Journal*, Vol. 12, No. 2, 2000.

- **Measurement:** This is the number of CEMs, capabilities against current and emerging threats—resulting in an increased number of CEMs and additional warfighting capacity.[13]

The experimentation and testing process is performed by the NIFC-CA System Engineering, Integration, and Test Integrated Product Team (IPT) using data from the pillar programs and other federated programs, such as the JSF, the F/A-18 E/F strike fighter, the E-2D Advanced Hawkeye program, and the AMOD program. Various experimental tests have been conducted over the life of the NIFC-CA project, including the following:

- September 2011: The Navy and the Army successfully executed a joint live-fire test to demonstrate NIFC-CA capability at White Sands Missile Range, New Mexico. This successful NIFC-CA test was the first live-fire engagement using tactical AEGIS baseline 9, CEC, SM-6, and JLENS systems and the first use of non-naval elevated sensor data to support an AEGIS OTH engagement.
- September 2012: The Navy executed its first live-fire demonstration to successfully test the integration of the F-35 with existing NIFC-CA architecture. An unmodified U.S. Marine Corps F-35B from the Marine Operational Test and Evaluation Squadron 1 was used as an elevated sensor to detect an OTH threat. The aircraft then sent data through its multifunction advanced data link to a ground station connected to USS Desert Ship (also known as LLS-1), a land-based launch facility designed to simulate a ship at sea. Employing the latest AEGIS weapon system baseline 9.C1 and a SM-6, the system successfully detected and engaged the target.
- From 2012 to 2019, multiple successful live-fire tests of the SM-6 against a variety of threats were executed using the ACB-9 through ACB-16 AEGIS Combat System baselines and the E-2D

[13] Jon Hill, "Systems Engineering Engagement Modes," Program Executive Office Integrated Warfare Systems briefing, Sea-Air-Space, May 17, 2016.

Advanced Hawkeye sensor. Through these live-fire tests, the Navy fielded NIFC-CA FTS Increment 1 capability in 2015 as a tactical option in fleet air defense.[14]

These examples highlight the process of experimentation and testing that the NIFC-CA project uses to overcome integration and interoperability problems that occur when combining large monolithic acquisition programs into an SoS capability. From these experiments and testing events, the Navy has solved the problem of how to integrate systems by understanding the quality of service needed to engage the specific threat.[15] The more difficult challenges are connectivity and quality of service, because these depend on the latencies and accuracies of the networks and, for specific threats, the reaction time needed to engage and destroy the threat. The challenge regarding reaction time includes not only the time involved in sending the data but also the network lag in getting the data to the shooter in time to engage the threat. Missile defense requires both exquisitely accurate data on the target and synchronous relationships between data systems, weapon-control systems, and flight-control systems because of the need to hit one missile moving at hundreds of miles per hour with another missile moving at hundreds of miles per hour. A minuscule error can mean a miss. The worst mass-casualty incident in the 1991 Gulf War—a Scud missile strike on Dhahran barracks that killed 28 Americans and wounded another 98—occurred because a software glitch caused a Patriot missile's timing to be 0.3433 seconds off.[16] This outcome shows the importance of experimentation and testing in complex kill chains.

Lessons Learned

NIFC-CA began with three exemplar kill chains and identified a standard for data structures early. This agreed-upon format and early con-

---

14 Robert F. Behler, *FY 2017 Annual Report*, Washington, D.C.: Office of the Director, Operational Test and Evaluation, January 2018.

15 Sydney J. Freedberg, Jr., "We CAN Tie Army, Navy Missile Defense Networks: Navy Experts," *Breaking Defense*, February 24, 2017.

16 Freedberg, 2017.

cept definition allowed the early use of experiments and ensured that all component programs were capable of understanding a common data format.

Also, the Navy conducted live-fire exercises over a period of more than 20 years to demonstrate the concepts of operations (CONOPS) and the component technologies, which led directly to the implementation of a pillar program (CEC). NIFC-CA leveraged underlying exercises for the component programs, adding a small number of well-defined internally funded test events. Thus, NIFC-CA leveraged the testing infrastructure and cost of the pillar programs rather than having to depend on the project office to create its own.

Naval Integrated Fire Control—Counter Air Project Documentation and Oversight

NIFC-CA is a systems engineering and test project rather than a formal acquisition program; therefore, the documentation that justifies and supports the project is not extensive. The majority of documents developed for the NIFC-CA project are technical and systems engineering documents. The traditional systems engineering V-diagram, in Figure 4.3, highlights the process and output (documents) for the NIFC-CA project. This diagram provides a detailed overview of how the systems engineering process works for the NIFC-CA project to address interoperability and integration issues that emerge from the four pillar programs.

The output documents from this process, as detailed in Figure 4.3, include the following:

- **NIFC-CA DoD Architecture Framework (DODAF):** an overarching, comprehensive architectural framework and conceptual model that facilitates the ability to make key decisions through organized information-sharing across DoD
- **A technical CONOP:** a bridge between the (often vague) program's or system's IOCs and the specific technical requirements needed to make it successful

Figure 4.3
Naval Integrated Fire Control—Counter Air Systems Engineering Processes and Products

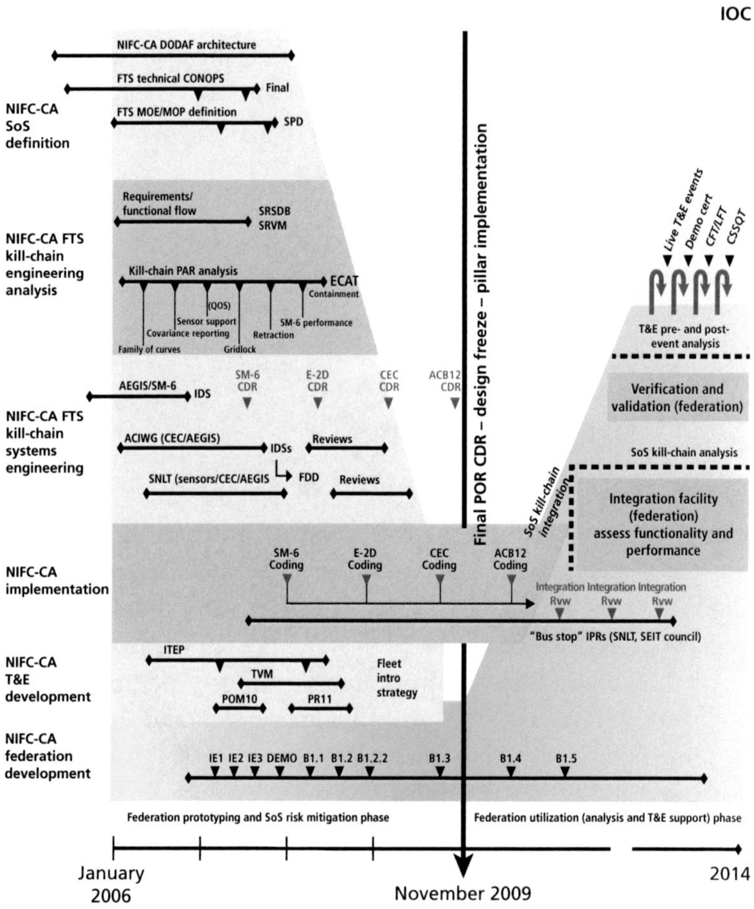

SOURCE: Adapted from McConnell and Jordan, 2013.
NOTE: ACIWG = AEGIS CEC Integration Working Group; CFT = cross-functional team; CSSQT = Combat Systems Ship Qualification Trials; ECAT = Expanded Counter Air Threat; FDD = full deployment decision; IDS = interface design specification; IPR = interim progress review; ITEP = Integrated Test and Evaluation Plan; LFT = Live-Fire Test; MOE = measure of effectiveness; MOP = measure of performance; PAR = performance assessment report; POM10 = Program Objective Memorandum for FY 2010; POR CDR = Program of Record, Critical Design Review; PR11 = Program Review for FY 2011; QOS = quality of service; Rvw = review; SEIT = Systems Engineering, Integration and Test; SNLT = Sensor Netting Leadership Team; SPD = system performance definition; SRSDB = System Requirements System Design Baseline; SRVM = specification requirements verification matrix; TVM = Traceability and Verification Matrix.

- **FTS Measures of Effectiveness:** a set of quantifiable measures used to define whether a system or architecture is effective for a given CONOP
- **Pillar program Critical Design Review:** a multidisciplinary technical review that assesses a system's final design using detailed specifications for each configuration item and produces a technical data package of each configuration item for the system's initial product baseline
- **Requirements functional flow analysis:** a functional analysis and allocation top-down process of translating system-level requirements into detailed functional and performance design criteria
- **Kill-chain engineering analysis:** the decomposition of a tightly integrated real-time kill chain and subsequent reallocation of that kill chain across independent component systems
- **Live-fire T&E events:** a test process that evaluates the vulnerability or lethality of a conventional weapon or weapon system.

The primary "programmatic" document for the NIFC-CA project is its Exhibit R-2, RDT&E Budget Item Justification.[17] This document is submitted annually with the Department of the Navy's budget submission to Congress to justify the funding request for the NIFC-CA project. This budget display provides a detailed summary of any project changes, budget or funding adjustments, accomplishments, and planned events. For example, the Integration and Test IPT tasks for FY 2018 included the following:

- two live-fire events and follow-on NIFC-CA battlespace assessment for AEGIS Baseline 9 (ACB 9) and SM-6 BLK 1
- White Sands Missile Range USS Desert Ship upgrades.

---

[17] Department of the Navy, 2018c.

The tasks for FY 2019 included the following:

- two live-fire events to verify the NIFC-CA 2019 capability improvements prior to fleet introduction in FY 2020 (one SM-6 Blk 1 regression test and one SM-6 Blk 1A)
- capability assessment from the commander of the Navy's Operational Test and Evaluation Force for NIFC-CA Increment 2 capability
- ongoing NIFC-CA Increment 3 risk-reduction testing.

Similar plans are outlined for modeling and simulation and engineering management and systems definition efforts. Figure 4.4 is the top-level planning schedule for the FTS kill chain that is also included in the R-2A budget display. The R-2A budget display also provides a detailed breakdown of the funding and government organizations and contractors executing the work in the three function areas: product development, T&E, and management services.

One area of contention among the Operational T&E community was the lack of a T&E master plan (TEMP) for the NIFC-CA project. The Operational T&E community stated that not having a NIFC-CA TEMP that specified operational tests in an environment that represents real warfighting scenarios was a serious deficiency and made it difficult to determine whether the NIFC-CA was operationally effective. Eventually, it was determined that the pillar programs would each have NIFC-CA live-fire test events in their program TEMPs rather than having a specific NIFC-CA TEMP.

The Director, Operational Test and Evaluation was responsible for ensuring that the Operational T&E community evaluated these NIFC-CA live-fire test events as part of the pillar acquisition programs. This approach removed the need to develop a complicated TEMP for the NIFC-CA project and focused the NIFC-CA pillar programs on operational suitability and effectiveness for NIFC-CA capabilities. This approach also enabled the NIFC-CA project to execute a capabilities-based acquisition project by levying minimal requirements onto the

Figure 4.4
Naval Integrated Fire Control—Counter Air From-the-Sea Planning Schedule

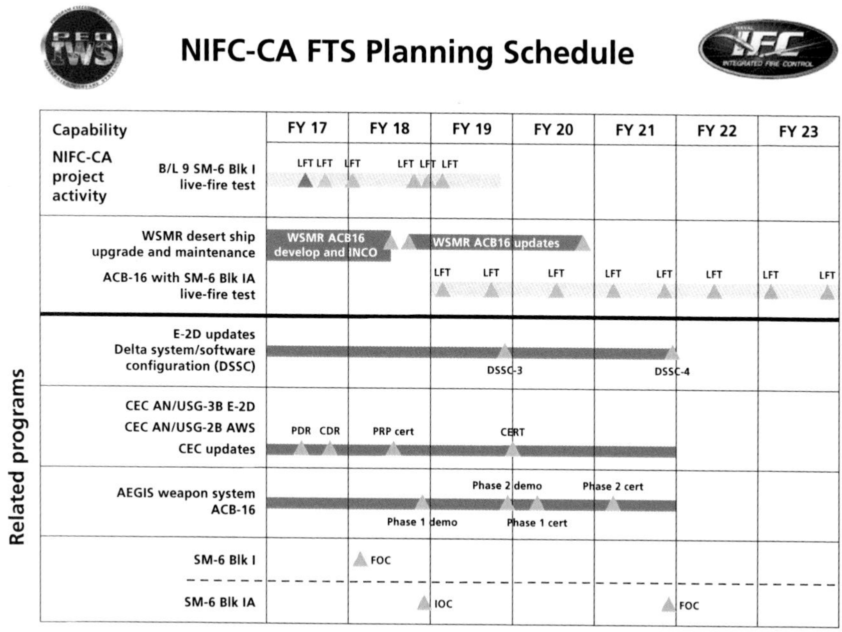

SOURCE: Adapted from Department of the Navy, 2018c.
NOTE: AWS = Aegis Weapon System; CERT = Certification; FOC = Full Operational Capability; INCO = Incorporation; PDR = Preliminary Design Review; PRP = Procurement Request Package; WSMR = White Sands Missile Range.

component systems while deriving SoS capability from the federation of these independent systems.[18]

Lessons Learned

Instead of a rigorous TEMP, which is typically required for major acquisition programs to rigorously test against predefined requirements, NIFC-CA approached their test events in a more scientific manner designed to test hypotheses. Test failure was seen as a learning opportunity rather than a project setback.

[18] Assistant Secretary of the Navy for Research, Development and Acquisition and Vice Chief of Naval Operations, 2002.

Naval Integrated Fire Control—Counter Air Development Timeline

Although the earliest test events related to NIFC-CA took place in 1994 and the operational requirement that eventually gave rise to NIFC-CA was first discussed formally in 1996,[19] the NIFC-CA project itself was not established until 2002.[20] Without analyzing the year-over-year spending or pace of tests, it is striking that the project had been in development for 13 years when NIFC-CA Increment 1 was fielded in March 2015 on the USS *Theodore Roosevelt* (CVN-71) CSG.[21] Although 13 years is a much shorter time period than the development timeline for a fifth-generation aircraft,[22] it is longer than the ten years needed to acquire the USS *Ford* (which was procured in FY 2008 and delivered in FY 2018, although it might not enter service until FY 2025).[23]

The long development time for NIFC-CA can be explained by two factors. First, at the inception of the NIFC-CA, the component programs were not complete, so part of the NIFC-CA process depended on major acquisition systems. For example, the E-2D Advanced Hawkeye declared IOC in 2014.[24] Notably, the CEC datalink that is critical to NIFC-CA's success achieved IOC in 1996, and the shipboard vari-

---

19 Undersecretary of Defense for Acquisition and Technology and Vice Chairman of the Joint Chiefs of Staff, 1999.

20 Assistant Secretary of the Navy for Research, Development and Acquisition and Vice Chief of Naval Operations, 2002.

21 Sam LaGrone, "Carrier Theodore Roosevelt Departs Norfolk for Middle East Deployment, New Homeport After Delay," USNI News, March 11, 2015.

22 The F-22 program began with the Advanced Tactical Fighter program requirements in 1981 and reached IOC in 2005 (24 years). The F-35 began in 1992 as the Joint Strike Fighter program and reached IOC in 2015 but is famously still in low-rate initial production as of 2019 (27 years).

23 Sam LaGrone, "Carrier Ford May Not Deploy Until 2024, 3rd Weapons Elevator Certified," USNI News, October 22, 2019; and Ronald O'Rourke, *Navy Ford (CVN-78) Class Aircraft Carrier Program: Background and Issues for Congress*, Washington, D.C.: Congressional Research Service, RS20643, updated August 14, 2020.

24 Majumdar, 2014.

ant achieved Full-Rate Production (FRP) in 2002, but the airborne variant for the E-2D was not FRP until 2014.[25] NIFC-CA's development timeline had to be aligned with these pillar programs; there was simply no way to deploy NIFC-CA before all of these components were ready.

A second factor likely responsible for the prolonged timeline was the decision to organize NIFC-CA as a project rather than as a traditional acquisition program. This decision had many benefits, as discussed earlier, but it also created a trade-off: Without the pressure of formalized requirements and regular milestone events and design reviews, the NIFC-CA project could proceed at its own pace. The Navy was careful to conduct detailed modeling and simulation on each incremental capability and did not assume missions or requirements that were deemed out of scope or unreasonable. This occurred to the chagrin of traditional acquisition T&E professionals, who regarded this freedom as license for the NIFC-CA management team to simply discard requirements they felt were too difficult or shift timelines to suit their needs.

Lesson Learned

NIFC-CA's lengthy development cycle was influenced by the acquisition timeline of its pillar programs but was also enabled by its organization as a project rather than as a program. The length of the development cycle was driven primarily by the need to integrate the legacy pillar programs into an integrated FoS, which they were not originally envisioned to support. Future increments of NIFC-CA will not require such long development cycles and are planned to be accomplished within much shorter timelines.

---

[25] DoD, *Cooperative Engagement Capability (CEC)*, Selected Acquisition Report (SAR), Washington, D.C., March 17, 2016.

Implications for Mosaic Warfare

This chapter has described the organizational construct, experimentation approach, and supporting documentation used in the development of NIFC-CA. Within the preceding sections, we have highlighted lessons that we believe are relevant to any attempts to acquire Mosaic warfare architecture and component systems. In this section, we reiterate these lessons learned and make observations regarding how they apply to Mosaic warfare, within the context of the recommendations of the Mitchell Report and the goals of Mosaic warfare, as outlined in Figure 4.1 and Table 4.1.

Potential Lessons Learned for Acquisition of Mosaic Warfare

The lessons learned from the U.S. Navy's implementation of NIFC-CA reveal potential benefits of a similarly structured mosaic system acquisition. First, the Navy's decision to structure NIFC-CA as a project with minimal (approximately $35 million per year) RDT&E funding for integration and interoperability of the pillar programs rather than to form a separate acquisition program with significant procurement funding resulted in a small increase in funding requirements on the underlying pillar programs of record. However, this decision also allowed NIFC-CA to survive the turbulence of budget cycles and reviews, including the cancellation of one of its core pillars (JLENS). A similar structure of leveraging existing programs of record, with redundant capabilities, in support of an integration *project* could allow DoD to focus mosaic acquisition efforts on the system architecture independently of efforts to acquire specific component systems. This idea also aligns with the Mitchell Report's recommendation to maintain existing force structures during transition.

Additionally, NIFC-CA began with three exemplar kill chains and identified a standard for data structures early. This agreed-upon format—and early concept definition—allowed the early use of experiments and ensured that all component programs were capable of understanding a common data format. If the mosaic vision can be similarly exemplified with a few pathfinder architectures, then efforts can be made to demonstrate the data structures and communications with experiments. This idea supports the Mitchell Report's recommendation to experiment on mosaic concepts and architectures.

Finally, the Navy conducted live-fire exercises over a period of more than 20 years to demonstrate the CONOPS and the component technologies, which led directly to the implementation of a pillar program (CEC). NIFC-CA leveraged underlying exercises for the component programs, adding a small number of well-defined test events, thus shifting most of the cost for testing onto the pillar programs rather than the NIFC-CA project office. If the mosaic project architecture is similarly designed, then it can similarly rely on component programs to fulfill most of its early experimentation. This supports the Mitchell Report's recommendations on maintaining an existing force structure, developing automated technology to share data across platforms, and conducting experimentation on mosaic concepts and architectures.

Developmental Hurdles for Mosaic Warfare

The lessons learned also expose potential challenges for the successful acquisition of a mosaic system. For example, NIFC-CA brought its distributed capability to bear by levying minimal T&E requirements onto component programs rather than executing an expansive TEMP. Although the mosaic project could follow a similar approach, this approach alone would likely be insufficient. Unlike NIFC-CA, Mosaic warfare depends heavily on dynamic behavior and coordination of loosely integrated systems and will include many more potential kill chains to test. DoD also must expend resources developing T&E strategies for AI systems, to provide some measure of assurance that systems will perform as expected in the field, as noted in the Mitchell Report's recommendations.[26] This is a monumental task and should not be overlooked. The solution might include careful design of the systems to provide some measure of confidence that behavior will not deviate from acceptable parameters, or it could include a life-cycle

---

[26] Although it does not fundamentally change the nature of this observation, the apparent need for advanced AI/ML to enable Mosaic warfare places increased importance on the need to develop a robust T&E strategy that can inspire confidence in how the system will behave—particularly if the policy implications surrounding the employment of lethal weapons without human operators in the loop are to be overcome.

monitoring approach in which systems are consistently monitored for responses to both real-world and synthetic inputs.[27]

The initial Mosaic warfare vision might take some time to field, particularly as new concepts, such as the goals of rapid composability and scalability to large numbers of platforms, are tested and built. This hurdle is analogous to the lengthy development time of NIFC-CA as its pillar programs progressed through their own acquisition programs. The challenges for Mosaic warfare might be more pronounced, because the vision for Mosaic warfare ultimately increases the diversity of systems and subsystems that will need to be assembled and the number of use cases that are envisioned. This is not necessarily a problem for Mosaic warfare, but success will depend on the ability to speed development for new capabilities once the architecture is acquired. Development of robust T&E strategies for AI systems, as noted in the Mitchell Report, will likely be a driving factor in this effort to speed acquisition of mosaic capabilities, particularly as the number of component systems and subsystems grows, leading to many possible ways in which the components can be arranged.

Conclusion

In this chapter, we have used the NIFC-CA project as an example development approach and analyzed the lessons learned to determine whether any of those lessons translate to the acquisition of a still-undefined Mosaic warfare system. We have highlighted several encouraging approaches for acquisition of Mosaic warfare, but there are many developmental hurdles that must first be surmounted, and a wholesale replication of the NIFC-CA development approach likely will fail to achieve all of the Mosaic warfare objectives.

Whatever lessons we can learn from NIFC-CA, one point is clear: This project followed a unique approach to development and fielding and largely owes its success to this fact. Although there can be no certainty that it would have failed otherwise, we believe that the chal-

---

27 See, for example, a recent paper on machine learning algorithm assessment and online monitoring: Eric Breck, Shanqing Cai, Eric Nielsen, Michael Salib, and D. Sculley, *The ML Test Score: A Rubric for ML Production Readiness and Technical Debt Reduction*, IEEE, 2017.

lenges NIFC-CA faced (primarily the cancellation of JLENS) would have been more problematic if NIFC-CA were structured as a major acquisition program. Implementing Mosaic warfare across the entire DoD enterprise will be a daunting task that is much more complex and challenging than was NIFC-CA, and although the NIFC-CA model is illuminating, there is more work to be done. The success of Mosaic warfare will likely require additional innovative approaches to acquisition.

CHAPTER FIVE

Recommendations

Our report has detailed two case studies that we conducted to make inferences about Mosaic warfare. The first case study examined the positive and negative responses of the immune system to pathogens. The second case study analyzed NIFC-CA to assess how Mosaic warfare could possibly be acquired. These case studies are presented in Chapters Three and Four, along with the series of observations each one generated. In this chapter, we collect those observations and distill them into policy recommendations focused on three areas: (1) programmatic recommendations, which concern how a Mosaic warfare acquisition should be structured; (2) R&D recommendations, which concern the development of necessary capabilities or technologies; and (3) T&E recommendations, which concern necessary changes or improvements to DoD's approach to testing new systems before they are deployed.

Some of these observations align closely with recommendations made by a DARPA-commissioned report on Mosaic warfare (the Mitchell Report, described earlier)—specifically, the need to develop automated technology for the transmission of information and new T&E approaches for AI and other adaptable programs and the need to conduct experimental verification of mosaic architectures and concepts.[1]

---

1 Deptula et al., 2019.

Programmatic Recommendations

The first set of recommendations pertains to the programmatic aspects of how a Mosaic warfare architecture should be acquired; they draw largely from our case study of NIFC-CA.

Identify Pathfinder Programs and Exemplar Vignettes Against Which to Develop the Mosaic Warfare Architecture

We noted that NIFC-CA began with three exemplar kill chains and identified a standard for data structures early. This agreed-upon format—and early concept definition—allowed the early use of experiments and ensured that all component programs were capable of understanding a common data format. If the mosaic vision can be similarly exemplified with a few pathfinder architectures, then efforts can be made to empirically demonstrate the data structures and communications.

Structure Acquisition of the Mosaic Warfare Architecture as a Project with Pillar Programs Rather than as a Program of Record

We noted that the decision to structure NIFC-CA as an integration project tied to multiple existing programs of record allowed it to survive the turbulence of budget cycles and reviews, including the cancellation of one of its core pillars (JLENS). A similar structure of leveraging existing programs of record, with redundant capabilities, in support of an integration *project* could allow DoD to focus mosaic acquisition efforts on the system architecture, independent of efforts to acquire specific component systems.

Research and Development Recommendations

The second set of recommendations pertains to new technology or processes that are needed to enable the prerequisites for a successful Mosaic warfare architecture; these recommendations are largely based on our immune system analogy.

Develop Algorithms for Robust Data Sharing and Distributed Processing Across Large Networks of Mobile Systems

We noted that immune systems function well in large part because they are adaptable. Thus, DoD should develop robust AI capabilities to automatically share relevant data and design responses based on observed adversary behaviors and capabilities.

Develop Low-Cost Rapid Manufacturing of Platforms and Payloads That Will Make Up the Mosaic Warfare System

We also noted that mosaic systems are often inefficient because of the redundant nature of their platforms and the swarming nature of the immune response. It is not clear whether this inefficiency will be a fundamental trait, but it appears tightly linked to the resilience noted earlier. It will be imperative that any fielded mosaic system reduces the cost of individual platforms as much as possible.

Develop Distributed Manufacturing Capabilities to Construct the Platforms and Payloads That Will Make Up the Mosaic Warfare System Close to Where They Are Needed

We noted that a rapid, tailored response could be provided through the use of distributed production facilities. Further R&D to mature additive manufacturing technologies and designs is critical to enabling this capability.

Develop Distributed and Automated C2 Algorithms and Processes That Can Translate High-Level Commands to Individual Platforms and Adapt Behavior to Changing Environments and Threat Behavior

Finally, we noted that mosaic systems likely will be difficult to control. This feature was borne out in modeling and simulation conducted for our study and reported elsewhere.[2] This difficulty underscores the

---

[2] Timothy R. Gulden, Jonathan Lamb, Jeff Hagen, and Nicholas A. O'Donoughue, *Modeling Rapidly Composable, Heterogeneous, and Fractionated Forces: Findings on Mosaic Warfare from an Agent-Based Model*, Santa Monica, Calif.: RAND Corporation, RR-4396-OSD, 2021.

need to develop robust algorithms that can handle the transmission of information and response, possibly through stigmergy.[3]

Test and Evaluation Recommendations

The third and final set of recommendations pertains to T&E procedures. These recommendations are drawn from both case studies.

Rely on Pillar Programs to Provide Early Test Opportunities, and Approach Each Test Event with a Willingness to Fail and a Learning Objective

We noted that NIFC-CA leveraged underlying exercises for the component programs, adding a small number of well-defined test events, thus shifting most of the cost for testing onto the pillar programs rather than the NIFC-CA project office. Instead of using a rigorous TEMP, which is typically required for major acquisition programs to rigorously test against predefined requirements, NIFC-CA approached its test events in a more scientific manner designed to test hypotheses. Test failure was seen as a learning opportunity rather than a project setback.

If the mosaic project architecture is similarly designed, then it can also rely on component programs to fulfill most of its early experimentation.

Develop an Approach to Test Adaptable Systems (Including Artificial Intelligence/Machine Learning Algorithms)

This approach might require the relaxation of reliability and robustness standards, with the understanding that mosaic systems are inherently more redundant. We noted that, unlike NIFC-CA, Mosaic warfare depends heavily on dynamic behavior and coordination of loosely integrated systems and will include many more potential kill chains to test. DoD also must expend resources developing T&E strategies

---

[3] *Stigmergy* is a "mechanism of indirect coordination in which the trace left by an action in a medium stimulates subsequent actions" (Francis Heylighen, "Stigmergy as a Universal Coordination Mechanism I: Definition and Components," *Cognitive Systems Research*, Vol. 38, June 2016, p. 4).

for AI systems, to provide some measure of assurance that systems will perform as expected in the field, as noted in the Mitchell Report's recommendations.[4]

The test plan for a Mosaic system might have to accept some likelihood of unexpected behavior and find some way to bound its risk of occurrence.

Test the Resilience of the Mosaic Warfare Architecture to Failure or Compromise of Component Systems

We noted that mosaic systems are resilient to individual failures; this resilience appears to be an innate trait but is worth explicit testing.[5] In the development of Mosaic warfare systems, experiments and simulations that involve random failure of a certain percentage of assets would help determine the robustness of a given architecture—in other words, how susceptible a given architecture is to failure as individual elements of that architecture fail.

Limitations in Our Analysis

In our report, we have used the immune response to foreign pathogens as an analogy to understand and critique mosaic concepts and have taken lessons learned from the NIFC-CA project for the acquisition of Mosaic warfare. As with all analogies, there are limitations.

The primary limitation in the immune response analogy is that the immune response is inherently a defensive scenario in which numerical superiority and proximity of forces are advantages. This model applies directly to base-defense scenarios but somewhat stresses

---

[4] Although it does not fundamentally change the nature of this observation, the apparent need for advanced AI/ML to enable Mosaic warfare places increased importance on the need to develop a robust T&E strategy that can inspire confidence in how the system will behave—particularly if the policy implications surrounding the employment of lethal weapons without human operators in the loop are to be overcome.

[5] The resilience of a mosaic force to individual platform failures is discussed in Justin Grana, Jonathan Lamb, and Nicholas A. O'Donoughue, *Findings on Mosaic Warfare from a Colonel Blotto Game*, Santa Monica, Calif.: RAND Corporation, RR-4397-OSD, 2021.

the application to force projection. Our comparisons, as drawn, are agnostic to these facts, but the underlying success of the immune system depends on the ability to muster large numbers of both general and specialized response forces and the ability to communicate information to reserve forces regarding the location and composition of adversary forces.

The primary limitation in our use of NIFC-CA as a case study is the fact that NIFC-CA, while distributed, lacks the dynamic kill-chain aspects that are central to Mosaic warfare. In the development and testing of NIFC-CA, tremendous care was taken to define and test the kill chains (from the E-2D or the AEGIS sensor to the F/A-18 strike aircraft or the SM-6). In contrast, Mosaic warfare is defined by the presence of multiple options for each element of the kill chain and the promise that some elements can be defined later and integrated into the architecture. This characteristic presents significant challenges for acquisition of the new systems and integration into the architecture.

References

Abramson, Neil, and Becky Melton, "Leukocytosis: Basics of Clinical Assessment," *American Family Physician*, Vol. 62, No. 9, November 2000, pp. 2053–2060.

Alberts, Bruce, Alexander Johnson, Julian Lewis, Martin Raff, Keith Roberts, and Peter Walter, *Molecular Biology of the Cell*, 4th ed., New York: Garland Science, 2002.

Andersen, Mads Hald, David Schrama, Per thor Straten, and Jürgen C. Becker, "Cytotoxic T Cells," *Journal of Investigative Dermatology*, Vol. 126, No. 1, January 2006, pp. 32–41.

Assistant Secretary of the Navy for Research, Development and Acquisition and Vice Chief of Naval Operations, "Updated Responsibilities for Management of Naval Integrated Fire Control—Counter Air (NIFC-CA)," joint memorandum, October 11, 2002.

Behler, Robert F., *FY 2017 Annual Report*, Washington, D.C.: Office of the Director, Operational Test and Evaluation, January 2018. As of February 6, 2020: https://www.dote.osd.mil/Portals/97/pub/reports/FY2017/other/2017DOTEAnnualReport.pdf?ver=2019-08-19-113553-787

Breck, Eric, Shanqing Cai, Eric Nielsen, Michael Salib, and D. Sculley, *The ML Test Score: A Rubric for ML Production Readiness and Technical Debt Reduction*, IEEE, 2017. As of September 1, 2020: https://research.google/pubs/pub46555/

Cameron, Mark J., and David J. Kelvin, *Cytokines, Chemokines and Their Receptors*, Austin, Texas: Landes Bioscience, 2000–2013. As of September 1, 2020: https://www.ncbi.nlm.nih.gov/books/NBK6294/

Chaikeeratisak, Vorrapon, Katrina Nguyen, Kanika Khanna, Axel F. Brilot, Marcella L. Erb, Joanna K. C. Coker, Anastasia Vavilina, Gerald L. Newton, Robert Buschauer, Kit Pogliano, Elizabeth Villa, David A. Agard, and Joe Pogliano, "Assembly of a Nucleus-Like Structure During Viral Replication in Bacteria," *Science*, Vol. 355, No. 6321, January 2017, pp. 194–197.

Chaplin, David D., "Overview of the Immune Response," *Journal of Allergy and Clinical Immunology*, Vol. 125, No. 2, Supp. 2, February 2010, pp. S3–S23.

Condotta, Stephanie A., and Martin J. Richer, "The Immune Battlefield: The Impact of Inflammatory Cytokines on CD8+ T-Cell Immunity," *PLOS Pathogens*, Vol. 13, No. 10, 2017.

Cutts, C. J., and J. R. Speakman, "Energy Savings in Formation Flight of Pink-Footed Geese," *Journal of Experimental Biology*, Vol. 189, No. 1, 1994, pp. 251–261.

DARPA—*See* Defense Advanced Research Projects Agency.

"DARPA Tiles Together a Vision of Mosaic Warfare," webpage, Defense Advanced Research Projects Agency, undated. As of September 1, 2020:
https://www.darpa.mil/work-with-us/darpa-tiles-together-a-vision-of-mosiac-warfare

Davidson, Philip S., *Statement of Admiral Philip S. Davidson, U.S. Navy Commander, U.S. Indo-Pacific Command Before the Senate Armed Services Committee on U.S. Indo-Pacific Command Posture*, testimony presented before the U.S. Senate Committee on Armed Services, Washington, D.C., February 12, 2019. As of September 1, 2020:
https://www.armed-services.senate.gov/imo/media/doc/Davidson_02-12-19.pdf

Defense Advanced Research Projects Agency, "DARPA Perspective on AI," webpage, undated. As of September 1, 2020:
https://www.darpa.mil/about-us/darpa-perspective-on-ai

Defense Manpower Data Center, "DoD Personnel, Workforce Reports & Publications," webpage, undated. As of September 18, 2020:
https://www.dmdc.osd.mil/appj/dwp/dwp_reports.jsp

Department of the Navy, *Department of Defense Fiscal Year (FY) 2019 Budget Estimates: Aircraft Procurement, Navy*, Justification Book Vol. 1, Budget Activity 01–04, Washington, D.C., February 2018a. As of October 2, 2020:
https://www.secnav.navy.mil/fmc/fmb/Documents/19pres/APN_BA1-4_BOOK.pdf

———, *Department of Defense Fiscal Year (FY) 2019 Budget Estimates: Research, Development, Test & Evaluation, Navy*, Justification Book Vol. 1, Budget Activities 1, 2, and 3, Washington, D.C., February 2018b. As of February 6, 2020:
https://www.secnav.navy.mil/fmc/fmb/Documents/19pres/RDTEN_BA1-3_BOOK.pdf

———, *Department of Defense Fiscal Year (FY) 2019 Budget Estimates: Research, Development, Test & Evaluation, Navy*, Justification Book Vol. 3, Budget Activity 5, Washington, D.C., February 2018c. As of October 2, 2020:
https://www.secnav.navy.mil/fmc/fmb/Documents/19pres/RDTEN_BA5_Book.pdf

———, *Department of Defense Fiscal Year (FY) 2019 Budget Estimates: Research, Development, Test & Evaluation, Navy*, Vol. 5, Budget Activity 7, Washington, D.C., February 2018d. As of February 6, 2020:
https://www.secnav.navy.mil/fmc/fmb/Documents/19pres/RDTEN_BA7_Book.pdf

———, *Department of Defense Fiscal Year (FY) 2019 Budget Estimates: Weapons Procurement, Navy*, Vol. 1, Washington, D.C., February 2018e. As of February 6, 2020:
https://www.secnav.navy.mil/fmc/fmb/Documents/19pres/WPN_Book.pdf

———, *Department of Defense Fiscal Year (FY) 2019 Budget Estimates: Aircraft Procurement, Navy*, Vol. 1, Washington, D.C., March 2019. As of February 6, 2020:
https://www.secnav.navy.mil/fmc/fmb/Documents/20pres/APN_BA1-4_BOOK.pdf

Deptula, David A., and Heather R. Penney, with Lawrence A. Stutzriem and Mark A. Gunzinger, *Restoring America's Military Competitiveness: Mosaic Warfare*, Arlington, Va.: Mitchell Institute for Aerospace Studies, September 2019. As of September 1, 2020:
https://www.mitchellaerospacepower.org/single-post/2019/09/10/Restoring-Americas-Military-Competitiveness-Mosaic-Warfare

Flajnik, Martin F., and Masanori Kasahara, "Origin and Evolution of the Adaptive Immune System: Genetic Events and Selective Pressures," *Nature Reviews Genetics*, Vol. 11, No. 1, 2010, pp. 47–59.

Franks, Alexis L., and Jill E. Slansky, "Multiple Associations Between a Broad Spectrum of Autoimmune Diseases, Chronic Inflammatory Diseases and Cancer," *Anticancer Research*, Vol. 32, No. 4, April 2012, pp. 1119–1136.

Freedberg, Sydney J., Jr., "We CAN Tie Army, Navy Missile Defense Networks: Navy Experts," *Breaking Defense*, February 24, 2017.

———, "Air Force ABMS: One Architecture to Rule Them All?" *Breaking Defense*, November 8, 2019.

Stephanie Forrest, Steven A. Hofmeyr, and Anil Somayaji, "Computer Immunology," *Communications of the ACM*, Vol. 40, No. 10, October 1997, pp. 88–96.

Garnier, Simon, Jacques Gautrais, and Guy Theraulaz, "The Biological Principles of Swarm Intelligence," *Swarm Intelligence*, Vol. 1, 2007, pp. 3–31.

Gershgorn, Dave, "Elon Musk's Artificial Intelligence Group Opens a 'Gym' to Train A.I.," *Popular Science*, April 27, 2016.

Godin, Jean-Guy J., "Antipredator Function of Shoaling in Teleost Fishes: A Selective Review," *Canadian Naturalist*, Vol. 113, No. 3, January 1986, pp. 241–250.

Grana, Justin, Jonathan Lamb, and Nicholas A. O'Donoughue, *Findings on Mosaic Warfare from a Colonel Blotto Game*, Santa Monica, Calif.: RAND Corporation, RR-4397-OSD, 2021.

Grayson, Tim, "Mosaic Warfare," briefing slides, Washington, D.C.: Defense Advanced Research Projects Agency, Strategic Technology Office, July 27, 2018. As of September 1, 2020:
https://www.darpa.mil/attachments/STO-Mosaic-Distro-A.pdf

Gulden, Timothy R., Jonathan Lamb, Jeff Hagen, and Nicholas A. O'Donoughue, *Modeling Rapidly Composable, Heterogeneous, and Fractionated Forces: Findings on Mosaic Warfare from an Agent-Based Model*, Santa Monica, Calif.: RAND Corporation, RR-4396-OSD, 2021.

Hedenström, Anders, "Adaptations to Migration in Birds: Behavioural Strategies, Morphology and Scaling Effects," *Philosophical Transactions of the Royal Society B*, Vol. 363, No. 1490, January 2008, pp. 287–299.

Heylighen, Francis, "Stigmergy as a Universal Coordination Mechanism I: Definition and Components," *Cognitive Systems Research*, Vol. 38, June 2016, pp. 4–13.

Hill, Jon, "Systems Engineering Engagement Modes," Program Executive Office Integrated Warfare Systems briefing, Sea-Air-Space, May 17, 2016.

Hoare, D. J., J. Krause, N. Peuhkuri, and Jean-Guy J. Godin, "Body Size and Shoaling in Fish," *Journal of Fish Biology*, Vol. 57, No. 6, 2000, pp. 1351–1366.

Hoffman, William, Fadi G. Lakkis, and Geetha Chalasani, "B Cells, Antibodies, and More," *Clinical Journal of the American Society of Nephrology*, Vol. 11, No. 1, January 2016, pp. 137–154.

Hopp, Ann-Katrin, Anne Rupp, and Veronika Lukacs-Kornek, "Self-Antigen Presentation by Dendritic Cells in Autoimmunity," *Frontiers in Immunology*, Vol. 5, No. 55, February 2014.

Janeway, Charles A., Jr., Paul Travers, Mark Walport, and Mark J. Shlomchik, *Immunobiology*, 5th ed., New York: Garland Science, 2001.

Johns Hopkins Medicine, "Allergies and the Immune System," webpage, undated. As of September 2, 2020:
https://www.hopkinsmedicine.org/health/conditions-and-diseases/allergies-and-the-immune-system

Joint Publication 3-10.1, *Joint Tactics, Techniques, and Procedures for Base Defense*, Washington, D.C.: Chairman of the Joint Chiefs of Staff, July 23, 1996.

Joint Publication 3-35, *Deployment and Redeployment Operations*, Washington, D.C.: Chairman of the Joint Chiefs of Staff, January 10, 2018.

Jones, Jimmy, "System of Systems Integration Technology and Experimentation (SoSITE)," webpage, Defense Advanced Research Projects Agency, undated. As of September 1, 2020:
https://www.darpa.mil/program/system-of-systems-integration-technology-and-experimentation

JP—*See* Joint Publication.

Judson, Jen, "Congress Nails Runaway Blimp's Coffin Shut," Defense News, May 27, 2016.

Juul, Svend, Joseph S. Pliskin, and Harvey V. Fineberg, "Variation and Information in White Blood Cell Differential Counts," *Medical Decision Making*, Vol. 4, No. 1, 1984, pp. 69–80.

Kawamoto, Hiroshi, and Nagahiro Minato, "Myeloid Cells," *International Journal of Biochemistry & Cell Biology*, Vol. 36, No. 8, August 2004, pp. 1374–1379.

Kumar, Vinay, Abul K. Abbas, and Nelson Fausto, eds., *Robbins and Cotran Pathologic Basis of Disease*, 7th ed., Philadelphia, Pa.: Saunders, 2004.

LaGrone, Sam, "Carrier Theodore Roosevelt Departs Norfolk for Middle East Deployment, New Homeport After Delay," USNI News, March 11, 2015.

———, "Carrier Ford May Not Deploy Until 2024, 3rd Weapons Elevator Certified," USNI News, October 22, 2019.

Lang, Sharon Watkins, "SMDC History: Army Mountain Top Experiment," webpage, U.S. Army, January 18, 2018. As of September 9, 2020:
https://www.army.mil/article/199235/smdc_history_army_mountain_top_experiment

Lewis, James, and Martin Fowler, "Microservices: A Definition of This New Architectural Term," blog post, martinfowler.com, March 24, 2014. As of January 22, 2020:
https://martinfowler.com/articles/microservices.html

Lexico.com, "Monolithic," webpage, undated. As of October 2, 2020:
https://www.lexico.com/definition/monolithic

Majumdar, Dave, "Navy Declares IOC for E-2D Advanced Hawkeye," USNI News, October 16, 2014.

Mayer, Andreas, Yaojun Zhang, Alan S. Perelson, and Ned S. Wingreen, "Regulation of T Cell Expansion by Antigen Presentation Dynamics," *Proceedings of the National Academy of Sciences of the United States of America*, Vol. 116, No. 13, March 2019.

Mazzei, Patricia, Thomas Gibbons-Neff, and Christine Hauser, "Trainee on Military Base Mounts Deadly Attack," *New York Times*, December 6, 2019.

McConnell, Jeffrey H., and Lorra L. Jordan, "Naval Integrated Fire Control—Counter Air Capability–Based System-of-Systems Engineering," *Leading Edge*, February 2013, pp. 24–31.

Mlot, Nathan J., Craig A. Tovey, and David L. Hu, "Fire Ants Self-Assemble into Waterproof Rafts to Survive Floods," *Proceedings of the National Academy of Sciences of the United States of America*, Vol. 108, No. 19, 2011, pp. 7669–7673.

Morton, John F., "The Aegis Warship: Joint Force Linchpin for IAMD and Access Control," *Joint Forces Quarterly*, Vol. 8, No. 1, January 1, 2016.

Niu, Hua, Qingming Xiong, Akitsugu Yamamoto, Mitsuko Hayashi-Nishino, and Yasuko Rikihisa, "Autophagosomes Induced by a Bacterial Beclin 1 Binding Protein Facilitate Obligatory Intracellular Infection," *Proceedings of the National Academy of Sciences of the United States of America*, Vol. 109, No. 51, December 18, 2012, pp. 20800–20807.

Ocko, Samuel A., and L. Mahadevan, "Collective Thermoregulation in Bee Clusters," *Journal of the Royal Society Interface*, Vol. 11, No. 91, 2014.

O'Rourke, Ronald, *Navy Ford (CVN-78) Class Aircraft Carrier Program: Background and Issues for Congress*, Washington, D.C.: Congressional Research Service, RS20643, updated August 14, 2020.

Pernin, Christopher G., Elliot Axelband, Jeffrey A. Drezner, Brian B. Dille, John Gordon IV, Bruce J. Held, K. Scott McMahon, Walter L. Perry, Christopher Rizzi, Akhil R. Shah, Peter A. Wilson, and Jerry M. Sollinger, *Lessons from the Army's Future Combat Systems Program*, Santa Monica, Calif.: RAND Corporation, MG-1206-A, 2012. As of January 14, 2020:
https://www.rand.org/pubs/monographs/MG1206.html

Phonekeo, Sulisay, Nathan Mlot, Daria Monaenkova, David L. Hu, and Craig Tovey, "Fire Ants Perpetually Rebuild Sinking Towers," *Royal Society Open Science*, Vol. 4, No. 7, July 2017.

Piot, Peter, Michael Bartos, Peter D. Ghys, Neff Walker, and Bernhard Schwartländer, "The Global Impact of HIV/AIDS," *Nature*, Vol. 410, April 2001, pp. 968–973.

Pitcher, Tony J., "Functions of Shoaling Behaviour in Teleosts," in Tony J. Pitcher, ed., *Behaviour of Teleost Fishes*, Boston, Mass.: Springer, 1986, pp. 294–337.

Program Executive Office Unmanned and Small Combatants, "LCS Mission Modules Program: Transition from Development to Production," briefing slides, Washington, D.C.: Naval Sea Systems Command, undated. As of January 22, 2020:
https://www.navsea.navy.mil/Portals/103/Documents/Exhibits/SNA2019/MM-Zobel.pdf?ver=2019-01-15-175909-530

Raje, Nikita, and Chitra Dinakar, "Overview of Immunodeficiency Disorders," *Immunology and Allergy Clinics of North America*, Vol. 35, No. 4, November 2015, pp. 599–623.

Ring, G. H., and F. G. Lakkis, "Breakdown of Self-Tolerance and the Pathogenesis of Autoimmunity," *Seminars in Nephrology*, Vol. 19, No. 1, January 1999, pp. 25–33.

Seeley, Thomas D., and P. Kirk Visscher, "Group Decision Making in Nest-Site Selection by Honey Bees," *Apidologie*, Vol. 35, No. 2, March–April 2004, pp. 101–116.

Shan, Amar, "Heterogeneous Processing: A Strategy for Augmenting Moore's Law," *Linux Journal*, January 2, 2006.

Shlapak, David A., and Michael W. Johnson, *Reinforcing Deterrence on NATO's Eastern Flank: Wargaming the Defense of the Baltics*, Santa Monica, Calif.: RAND Corporation, RR-1253-A, 2016. As of February 6, 2020: https://www.rand.org/pubs/research_reports/RR1253.html

Silmon de Monerri, Natalie C., and Kami Kim, "Pathogens Hijack the Epigenome: A New Twist on Host-Pathogen Interactions," *American Journal of Pathology*, Vol. 184, No. 4, April 2014, pp. 897–911.

Syring, Jim, "Navy IAMD Capabilities," briefing slides for NDIA State of IAMD Symposium, Washington, D.C.: Program Executive Office, Integrated Warfare Systems, July 12, 2012. As of September 1, 2020: https://ndiastorage.blob.core.usgovcloudapi.net/ndia/2012/IAMD/Syring.pdf

Tai, Yu, Qingtong Wang, Heinrich Korner, Lingling Zhang, and Wei Wei, "Molecular Mechanisms of T Cells Activation by Dendritic Cells in Autoimmune Diseases," *Frontiers in Pharmacology*, Vol. 9, No. 642, June 2018.

Tirpak, John A., "Find, Fix, Track, Target, Engage, Assess," *Air Force Magazine*, July 1, 2000.

Undersecretary of Defense for Acquisition and Technology and Vice Chairman of the Joint Chiefs of Staff, "Land Attack Cruise Missile Defense (CMD)," joint memorandum, Washington, D.C., January 11, 1999.

Upton, Lee O., and Lewis A. Thurman, "Radars for the Detection and Tracking of Cruise Missiles," *Lincoln Laboratory Journal*, Vol. 12, No. 2, 2000, pp. 355–366.

U.S. Air Force, "Dynamic Targeting and the Tasking Process," Annex 3-60, Maxwell Air Force Base, Ala.: Curtis E. LeMay Center for Doctrine Development and Education, last updated March 15, 2019. As of September 19, 2019: https://www.doctrine.af.mil/Portals/61/documents/Annex_3-60/3-60-D17-Target-Dynamic-Task.pdf

U.S. Bureau of the Census, "Chapter D: Labor," *Historical Statistics of the United States: Colonial Times to 1970*, Part 1, Washington, D.C.: U.S. Department of Commerce, 1975, pp. 121–182. As of September 18, 2020:
https://www2.census.gov/library/publications/1975/compendia/hist_stats_colonial-1970/hist_stats_colonial-1970p1-chD.pdf

U.S. Department of Defense, *Cooperative Engagement Capability (CEC)*, Selected Acquisition Report (SAR), Washington, D.C., March 17, 2016. As of September 1, 2020:
https://apps.dtic.mil/sti/pdfs/AD1019001.pdf

U.S. Office of Management and Budget, "Historical Tables," webpage, undated. As of September 18, 2020:
https://www.whitehouse.gov/omb/historical-tables//

Valenta, Rudolf, "The Future of Antigen-Specific Immunotherapy of Allergy," *Nature Reviews Immunology*, Vol. 2, June 2002, pp. 446–453.

Vincent, Julian F. V., Olga A. Bogatyreva, Nikolaj R. Bogatyrev, Adrian Bowyer, and Anja-Karina Pahl, "Biomimetics: Its Practice and Theory," *Journal of the Royal Society Interface*, Vol. 3, No. 9, 2006, pp. 471–482.

Vitetta, Ellen S., Michael T. Berton, Christa Burger, Michael Kepron, William T. Lee, and Xiao-Ming Yin, "Memory B and T Cells," *Annual Review of Immunology*, Vol. 9, 1991, pp. 193–217.

Vivier, Eric, David H. Raulet, Alessandro Moretta, Michael A. Caligiuri, Laurence Zitvogel, Lewis L. Lanier, Wayne M. Yokoyama, and Sophie Ugolini, "Innate or Adaptive Immunity? The Example of Natural Killer Cells," *Science*, Vol. 331, No. 6013, January 2011, pp. 44–49.

Wherry, E. John, and David Masopust, "Adaptive Immunity: Neutralizing, Eliminating, and Remembering for the Next Time," in Michael G. Katze, Marcus J. Korth, G. Lynn Law, and Neal Nathanson, eds., *Viral Pathogenesis*, 3rd ed., London, UK: Elsevier, Academic Press, 2016, pp. 57–69.

Wilson, Clay, *Network Centric Operations: Background and Oversight Issues for Congress*, Washington, D.C.: Congressional Research Service, March 15, 2007. As of September 1, 2020:
https://fas.org/sgp/crs/natsec/RL32411.pdf

Xie, Jianming, Cristina M. Tato, and Mark M. Davis, "How the Immune System Talks to Itself: The Varied Role of Synapses," *Immunological Reviews*, Vol. 251, No. 1, January 2013, pp. 65–79.

Yen, Yung-Feng, Pei-Hung Chuang, I-An Jen, Marcelo Chen, Yu-Ching Lan, Yen-Ling Liu, Yun Lee, Yen-Hsu Chen, and Yi-Ming Arthur Chen, "Incidence of Autoimmune Diseases in a Nationwide HIV/AIDS Patient Cohort in Taiwan, 2000–2012," *Annals of the Rheumatic Diseases*, Vol. 76, No. 4, April 2017, pp. 661–665.